Mobility Engineering

Mahesh Chandra Tandon
Purnendu Ghosh
Editors

Mobility Engineering

Proceedings of CAETS 2015 Convocation on Pathways to Sustainability

Springer

Editors
Mahesh Chandra Tandon
Indian National Academy of Engineering
Gurgaon, Haryana
India

Purnendu Ghosh
Birla Institute of Scientific Research
Jaipur, Rajasthan
India

ISBN 978-981-10-3098-7 ISBN 978-981-10-3099-4 (eBook)
DOI 10.1007/978-981-10-3099-4

Library of Congress Control Number: 2016957698

Printed on acid-free paper

This Springer imprint is published by Springer Nature
The registered company is Springer Nature Singapore Pte Ltd.
The registered company address is: 152 Beach Road, #22-06/08 Gateway East, Singapore 189721, Singapore

Foreword

I am delighted that the Post Proceedings of the International Council of Academies of Engineering and Technological Sciences (CAETS) 2015 Convocation on "**Pathways to Sustainability: Energy, Mobility and Healthcare Engineering**" held in New Delhi, India on Oct 13–14, 2015 are being brought out. CAETS, being a non-governmental international organization of Academies of Engineering from 26 countries, which has the broad objective of advising governments and providing an international form for addressing all issues of concern pertaining to engineering and technology, plays a vital role in promotion of the engineering profession. Indian National Academy of Engineering (INAE), a member academy of CAETS, is proud to have had the opportunity of hosting the CAETS 2015 Annual Meeting and Convocation at New Delhi, India.

The CAETS event brought together eminent engineering luminaries and bright young engineers from 24 countries on a single platform who shared their country-specific knowledge on the themes of the Convocation. The CAETS 2015 Convocation focussed on the topics of energy, mobility and healthcare engineering which are of topical interest all over the world in recent times. Energy security, provision of economical modes of transport and mobility and affordable and high-quality facilities for healthcare are necessary for the citizens of the world in order to lead comfortable and secure lives. It is indeed apt that the Convocation addressed key issues in the fields of energy, mobility and healthcare engineering that play a vital role in the lives of mankind.

It is a matter of pride that the Convocation had representation of eminent engineers from academia, R&D and industry from over 24 countries from across the globe who shared their valuable experience and expertise and enriched the deliberations with their thought-provoking presentations. These presentations were indeed brilliant and stimulating and have resulted in interesting papers which are showcased in the Post Proceedings. I thank all the authors for devoting their precious time for contributing papers for the proceedings.

I express my sincere thanks to the editors of the CAETS 2015 Convocation Post Proceedings for their untiring efforts at every stage of planning of the event and in publication of the Post Proceedings. The assistance and initiative taken by Springer in bringing out the publication are deeply appreciated.

On behalf of Indian National Academy of Engineering (INAE), it is my proud privilege to express my deepest gratitude to the CAETS organization for giving us an opportunity to host this event of international repute for the first time in India.

I sincerely hope that the event left many pleasant and memorable memories for all the delegates and am certain that the Post Proceedings of the CAETS 2015 Convocation on "Pathways to Sustainability: Energy, Mobility and Healthcare Engineering" shall be well accepted in the engineering realm and by the engineering community.

Dr. B.N. Suresh
President, INAE

Preface

Preamble

The CAETS Convocation 2015, hosted by the Indian National Academy of Engineering (INAE), was held on October 13–14, 2015, at New Delhi, India. The CAETS Governing Council had chosen the theme on sustainable engineering pathways for energy, mobility and healthcare sectors for the conference attended by 250^{+} delegates comprising of the Fellowship of CAETS member academies, senior engineering experts, scientists, thought leaders, policy makers and industry executives from Belgium, Canada, China, France, Germany, Hungary, India, Japan, Netherlands, South Africa, Spain, Sweden, Switzerland, UK, USA, Uruguay and other countries. This international event was formally inaugurated by Dr. Harsh Vardhan, Hon'ble Minister of Science and Technology, Government of India, and he assured that the recommendations of the CAETS Convocation will be acted upon by the government. Professor Asutosh Sharma, Secretary, Department of Science and Technology, Government of India addressed the gathering and highlighted the importance of the selected themes to developing nations.

The two unique features of CAETS 2015 were the special gesture made by the INAE (a) to provide 50% international travel support and complimentary accommodation to encourage young engineers below 45 years of age to deliver lectures in the technical sessions along with eminent senior engineers and (b) to organize poster sessions and awards to encourage very young engineers and research scholars from CAETS member academies to present their research work. These initiatives paid rich dividends by attracting the participation of 30^{+} young engineers in CAETS Convocation. Another major initiative is the formal launching of the electronic version of the CAETS report on "Transitioning to Lower Carbon Economy: Technological and Engineering Considerations for Building and Transportation Sectors" during the inaugural function of CAETS Convocation-2015 by the chief guest. This is a joint initiative of 12 CAETS member academies from Australia, Canada, China, Germany, India, Japan, Korea, South Africa, Switzerland and UK. The experts from these academies met at Capetown (South Africa), Beijing (China),

Berlin (Germany) and London (UK) during 2013–15 to identify technological and engineering priorities and emerging options for transition to lower carbon economy in building and transportation sectors.

Recommendations

Energy

Global energy consumption will grow from 524 to 820 quadrillion BTU between 2010 and 2040 leading to GHG emission increase by more than 40% from the present level under business as usual scenario. Since the dynamics of energy growth is driven by the population and wealth enhancements, it is important to keep track of these growth dynamics in different countries. Energy resources sustainability as well as their cost efficiencies would dictate as rapid shift as possible to lower carbon non-fossil energy systems. Since under this context, recent developments in three-stage thorium-based nuclear energy, large capacity solar thermal power plants, fuel cell-based independent propulsion systems for strategic and civilian applications, wind generated hydrogen as a renewable energy carrier and unconventional gas from coal seams, underground coal gasifications and biohydrogen from waste resources provide potential lower carbon energy options. A joint voluntary initiative by CAETS member academies from Argentina, Australia, Canada, China, Germany, South Africa, Switzerland and UK on unconventional gas involving basic and social scientists, engineers and economists on unconventional gas provides a new model for CAETS member academies to emulate in the future.

Developing energy scenarios under various application modes is a powerful mechanism for devising future energy efficient systems. A typical example is the highly interconnected electricity, transport and heating sectors. The coal-based energy continues to attract engineers to develop cleaner process options and minimize GHG emissions keeping in mind the transitional priorities. The actions include upgradation of beneficiation, ultra supercritical combustion and integrated gas combined cycle concept for gasification, co-firing of coal and biomass; China's achievements in some of these areas are noteworthy. In energy materials, in which significant developments are taking place worldwide, we have to look beyond silicon for next generation solar cells by adopting conjugate organics, inorganic quantum dots and mixed semiconductor oxides/peroxides. Similar challenges are foreseen in high-energy-density battery and membrane materials for fuel cells. Light-based technologies can play a major part in developing future energy efficient systems. There is a need to integrate photonic and biophotonic technologies. In this international year of light technologies, such initiatives have to be taken to achieve multi-functionality, tunability, low power for fast operations, reconfigurability and cost-effectiveness so much necessary for energy, mobility and healthcare sectors.

The transition to lower carbon regime requires a careful assessment of current energy and emission loads at national level for setting energy/emission reduction targets and timelines. Selection of cost-effective technology and engineering options and their incentivization for creating new markets are the major challenges. The expert panel, which deliberated on these issues, stressed the need to develop high-strength lightweight materials including nanocomposites and nanostructured steels and bio, electrochemical and catalytic options for hydrogen generation adaptable for building and transportation sectors. CAETS need to work closely with related international bodies such as World Economic Forum, International Panel on Climate Change and allied agencies.

Mobility

The future engineering challenges lie in the development of semi-automated and fully automated transport vehicles, intelligent driver–vehicle interface and innovative multi-axle hydraulic trailers for heavy loads. The fast-changing digital technology has opened up new avenues for electrification of rail, air and marine transport vehicles. The five digital forces, viz. cloud computing, mobile technologies, social networks, big data and robotics will make high impact on these developments.

Exciting developments are taking place in bridge design and construction technologies with urban and rural transport restructuring, virtual mobility and carbon footprint minimization during their construction as defining factors. Novel retrofit technologies are needed for their restoration. A life cycle approach embracing bridge design, construction, maintenance, restoration and dismantling is a future priority area for civil engineers.

The recent developments in rapid urban rail transportation systems in China, India and Japan have demonstrated the new engineering skills acquired in planning, design and execution of underground tunnels and structures. From material engineering point of view, material recycle and functionalized material applications are high-priority areas in evolving sustainable growth.

The dilemmas in mass transportation, in emerging economies are many. It is important to consider seamless connectivity, smart mobility, enforcement of advanced safety measures, equitable allocations of road space for multiple types of vehicles and time-variant traffic demands in urban planning and design.

Health Care

Multitude of engineering challenges await healthcare sector in terms of new diagnostic tools, next generation medical devices and application of informatics and analytics. The recent advances in nano and point-of-care diagnostic, scalable medical and remote neonatal monitoring systems have enhanced the chances of early and accessible diagnosis. Structural process concept has greatly helped in developing groundbreaking technologies for these systems in advanced nations with the help of multidisciplinary teams comprising engineers, product designers, business analysts and clinicians.

The concept of regenerative engineering which combines tissue engineering, material science, cell physics and developmental biology has enabled the technologies for next generation medical devices. Groundbreaking bionic ear and eye technologies have enabled notable progress against hearing and visual impairments. The recent advances in sensors, telecommunication, mobility engineering will play a major role in evolving next generation device technologies for both communicable and non-communicable diseases. Sensing and data analytic skills provide new transformational material opportunities in intensive health care. Advanced computer-aided tools based on big data analytics are needed for biologically meaningful insights into the enormous volume of microbiome data generated from sequencing platforms.

A panel of experts examined the issue of convergence of engineering and healthcare sciences. They are certain that the convergence is occurring impressively due to application of ICT and big analytics in healthcare systems. Several key issues including affordability of well-engineered systems in rural and urban environments and engineers' role in new drug discovery came up for discussion.

The major takeaways from CAETS Convocation are many. Achieving long-term sustainability under business as usual mode is virtually impossible in energy, mobility and healthcare sectors. Major emphasis has to be placed on energy consideration, expanded use of non-fossil primary energy, decarbonization of existing energy sources and enhancing the energy efficiency of individual systems. Higher investment in R&D and demonstration is essential for technologies which are close to market maturity and those requiring scale up. Commitment from policy makers, different stakeholders, academic community and market leaders is very essential for the commercial realization. Ethical practices in engineering are vital for achieving high level of successes in every application field particularly so in creating human artefacts that are non-existent in nature. Gender enhancement in engineering design endeavours is essential for developing rational systems. Younger engineers have to be given increased responsibilities in evolving innovative engineering systems. The demand for novel engineering solutions will increase exponentially while developing smart energy, mobility and healthcare systems. Research, prototyping and technology transfers in such systems have to be path-breaking in nature. The CAETS Engineering Community is committed to create a conducive environment for open access information sharing in the above sectors between the member academies.

Gurgaon, India Mahesh Chandra Tandon
Jaipur, India Purnendu Ghosh

Contents

About the Editors

Mahesh Chandra Tandon is an international expert in the field of structural engineering. Many of the structures designed by Prof. Tandon and his firm Tandon Consultants Pvt. Ltd. (TCPL) have been widely acclaimed and have received recognition in India as well as internationally. He is the president of Indian Association of Structural Engineers (2015–16) and the president of Indian Society of Wind Engineering (2015–16). He was conferred the Honorary Fellowship of the Indian Concrete Institute (2015). TCPL was given the Industry Excellence Award by the Institution of Engineers at the Indian Engineering Congress in 2015. Under the AICTE-INAE programme, he was appointed distinguished visiting professor at IIT Kanpur, IIT Roorkee and IIT Gandhinagar (2005–2015).

He has to his credit a number of major design projects in Prestressed Concrete, Reinforced Concrete and Structural Steel in India, Malaysia, Thailand, Russia and several countries in the Middle East, South East Asia and Europe. These projects include major bridges, long-span roofs, tall buildings and monumental structures. His projects employ state-of-the-art technologies and the latest available techniques in design and construction. His structures have been described as highly creative, futuristic, aesthetic, environmentally sensitive and innovative. He has spearheaded the development of many codes of practice on concrete and on bridge structures in India.

Professor Tandon has made significant contributions in the development of a culture for innovation in structural engineering both within and outside his organization by sharing his expertise and experience. His special areas of interest also include motivating the next generation to adopt civil engineering as their profession and vocation in life.

Purnendu Ghosh is the executive director of Birla Institute of Scientific Research, Jaipur. Prior to this, he was professor at Indian Institute of Technology Delhi. He worked at the Swiss Federal Institute of Technology (ETH), Zurich, Switzerland; University of Melbourne, Australia; National Research Centre for Biotechnology (GBF), Braunschweig, Germany. He has taken a lead role in augmenting bioprocess engineering and biotechnology activities in the country as a member of several task forces of the Department of Biotechnology, Government of India. He is currently the chief editor of Publications of Indian National Academy of Engineering. He is a columnist of a leading newspaper group. Besides technical publications, he has written books of different genres.

Renaissance and Electric Vehicles Development

C.C. Chan

Abstract This paper presents the global state of the art of electric vehicles. The engineering philosophy of electric vehicles development, the key features of different types of electric vehicles, the propulsion technology, recent development of batteries technology and commercialization road map are discussed.

Keywords Electric vehicle · Hybrid electric vehicle

1 Introduction

To contend with the existing multifaceted economical, societal and technical encounters in the new epoch, being open-minded is extremely vital. Namely, incorporate the eastern ancient civilization and the western modern civilization, and uphold the spirit of renaissance. The dawn of the commercialization of electric vehicles (EVs) is coming closely. Nevertheless, the remaining darkness still should be passed through to scope the morning mist and the bright sunshine. Undoubtedly, EVs will become a paradigm in the new era. The success of commercialization of EVs depends on the satisfactory tackling of four factors: initial cost, convenience of use, energy consumption and exhaust emission. For that reason, further efforts should be made towards the following three fundamental enablers or three goodness factors:

C.C. Chan (✉)
Chinese Academy of Engineering, Beijing, China
e-mail: ccchan@eee.hku.hk

C.C. Chan
Royal Academy of Engineering, London, UK

C.C. Chan
Hong Kong Academy of Engineering Sciences, Wan Chai, Hong Kong

C.C. Chan
World Electric Vehicles Association, Phoenix, USA

M.C. Tandon and P. Ghosh (eds.), *Mobility Engineering*,
DOI 10.1007/978-981-10-3099-4_1

1. Availability of excellent performance products at affordable cost;
2. Availability of excellent infrastructures that is efficient and friendly to use;
3. Availability of excellent business model to leverage the cost of barrier.

In this connection, the three significant points including the policy, industry and market should be incorporated. Meanwhile, professional collaborations between the key players in the automobile industry and the electric power industry are in demand. Electric vehicle industry can be a disruptive industry, since some great differences exist between the function, production and commercial chain of EVs and the conventional vehicles. In terms of function, electric vehicle is not only just a transportation media, but also an electric device with moving energy storage capability. Thus the integration of EVs and smart gird, of EVs and information and communication technologies, is pretty indispensable. Such integration and collaboration should aim at gradually achieving the common goal of four zeros: zero emission, zero gasoline, zero traffic accident, and zero traffic jam. In this paper, the global state of the art of electric vehicles, their key technologies and their infrastructure will be addressed.

2 Renaissance Scientists and Engineers

The Renaissance was a cultural movement that spanned the period roughly from the 14th to the 17th century. The author proposes to espouse the core essence of renaissance [2]. Renaissance scientists and engineers are characterized by not only understanding Why and How things work but also on Why and How the World works. The charisma of renaissance scientists and engineers can be summarized as follows:

- Global thinking rather than local thinking;
- Circular thinking rather than linear thinking;
- Closed loop thinking rather than open loop thinking;
- Life cycle thinking rather than partial life thinking.

The author also proposes to uphold the spirit of Chinese ancient philosophy of Yi-jing. Namely, the principles of changing are balance, holistic, dialectical, periodicity and unity of opposites.

3 Features of Electric Vehicles, Hybrid Electric Vehicles and Fuel Cell Vehicles

Governments recently have developed more strict limitations on fuel consumption and emissions. Since Battery Powered Electric Vehicles (BEVs) consume zero oil and produce zero emission, etc. they are considered as the best solutions to deal with energy crisis and global warming. Notwithstanding, high initial cost, long charging time, and short driving range have restricted the wide applications of BEVs. To tackle this, better infrastructures should be provided, which include the combination of slow, medium, fast, battery leasing, swapping and inductive charging, as well as the integration with smart grid.

Hybrid Electric Vehicles (HEVs) emerge to overcome all the drawbacks of Internal Combustion Engine (ICE) vehicles and BEVs. A HEV is composed of an ICE propulsion system and an electric machine (EM), and has an electric energy system. The driving range of a HEV is longer than a BEV. Compared with conventional ICE vehicles, HEVs are more economical in fuel. When an ICE vehicle stops, the ICE will be stopped. The efficiency of the ICE can be optimized by an electric drive system. Hence, oil consumption and emission can be reduced. During braking and down slope driving modes, the kinetic energy can be recovered into the energy system. When HEVs operate in electric mode, they can be driven silently with zero-emission in a certain range. If the battery can be recharged through a plug connecting to an electric source, this range can be enlarged. HEV like this is called Plug-in Hybrid Electric Vehicle (PHEV). Moveover, the onboard EVs of HEVs can promote the controllability of vehicle control, such as vehicle stability control (VSC) and antilock braking (ABS), thus providing better performance.

The electricity is generated from hydrogen and air by fuel cells in Fuel Cell Vehicles (FCVs). It can be used to drive the vehicle or stored in an electricity storage system including battery parks or supercapacitors. Since FCVs only emit water vapour, they can have higher efficiency. Now there are three major issues related to FCVs. The first one is high price and the life cycle problem of fuel cells. Secondly, the energy density of hydrogen onboard storage needs to be improved. Thirdly, it is required to construct more hydrogen distribution and refueling infrastructure.

4 Engineering Philosophy of Electric Vehicle Development

The integration of automobile engineering and electrical engineering is the fundamental philosophy of the overall EV engineering. Hence, system the integration and optimization can be the primary approach to balancing the EV performances and cost. The features of electric propulsion essentially differ from those of engine propulsion. Under this circumstance, novel design approaches are critical for EV engineering. Furthermore, advanced energy resources and smart energy management are the key enabling factors for EV to compete with ICEVs. From the marketable perspective of EVs, the overall cost effectiveness will be the vital factor. In the whole, system integration and optimization is the core idea of engineering philosophy.

5 EV Propulsion

The vital component of the EVs is the electric motor drive. Its function is to link power source (such as batteries) with vehicle wheels, achieving the energy and information transmission in either direction as needed, with simultaneously and continuously driver response and high efficiency control. Therefore, as the kernel technology for EVs, HEVs and FCVs, the electric motor drive plays a significant role in the EVs components. The principal requirements of the traction motor drive are the following:

1. Power density and high torque density;
2. Quite wide speed range, including constant-power and constant-torque regions;
3. High efficiency coverage in large torque and speed ranges;
4. High torque for low-speed starting and climbing, and high power for high-speed cruising;
5. Rapid torque response;
6. High intermittent overload capability for overtaking;
7. High reliability and robustness for vehicular environment;
8. Relatively low acoustic noise;
9. Acceptable cost.

Figure 1 illustrates the typical torque/power-speed characteristics, which is necessary for traction motor drives. In addition to have good electromagnetic design to satisfy the torque-speed characteristic requirement, it is also essential to have superior mechanical design with unique stress analysis to satisfy sufficient motor mechanical strength under vehicles severe operating condition, and excellent cooling design with unique fluid analysis to satisfy the motor should not over heat under vehicle severe operating condition.

In order to meet these particular requirements, when determining the power rating and torque-speed specification of the motor drive, it's compulsory to consider the driving cycles and system-level. Novel control strategies and motor design

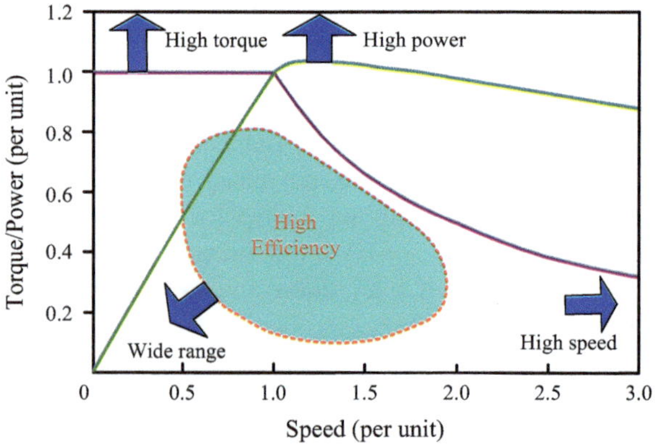

Fig. 1 Torque/Power requirements for traction motors

technologies are hot pursuits for the sake of achieving the speed range extension, the system efficiency optimization and the high-efficiency region enlargement. And by adopting the advanced electronic products, the system performance is improved while the total cost is reduced in the meantime. Figure 2 shows the functional block diagram of a motor drive for EVs.

With the rapid technological development of power electronics, microelectronics, electric motors and control strategies, the various selectivity of motor drives for EVs is becoming possible. Due to the advantages of achieving high torque at low speeds and easy control, DC motor drives have been utilized for EV propulsion traditionally. However, it's inconvenient for DC motor usage because of careful maintenance is necessary caused by its commutator and brushes. Contemporary technological growth have made the alternative of DC motors drives by other advanced motor drives possible, and the later options possess definite advantages such as high power density, robust, reliable, high efficiency, efficient regenerative braking, and maintenance free and so on. Amidst them, the highest popularity and most mature one belongs to the vector controlled induction motor drive, in spite of performing relatively low efficiency at light-load ranges. In addition, permanent magnet (PM) brushless motors are endowed with the highest power density and efficiency over the others, though undergoing a trouble in the constant-power high-speed region to achieve flux weakening control. As a special branch of the main stem of the PM brushless motors, the PM hybrid motor possess an integrated air-gap flux composing of PM flux and field-winding flux, which is supplied by an auxiliary DC field winding incorporated skillfully. The flexible air-gap flux can be achieved by regulating the field-winding excitation current, hence even in a wide speed range, the optimal efficiency is also available. Besides, benefiting of the

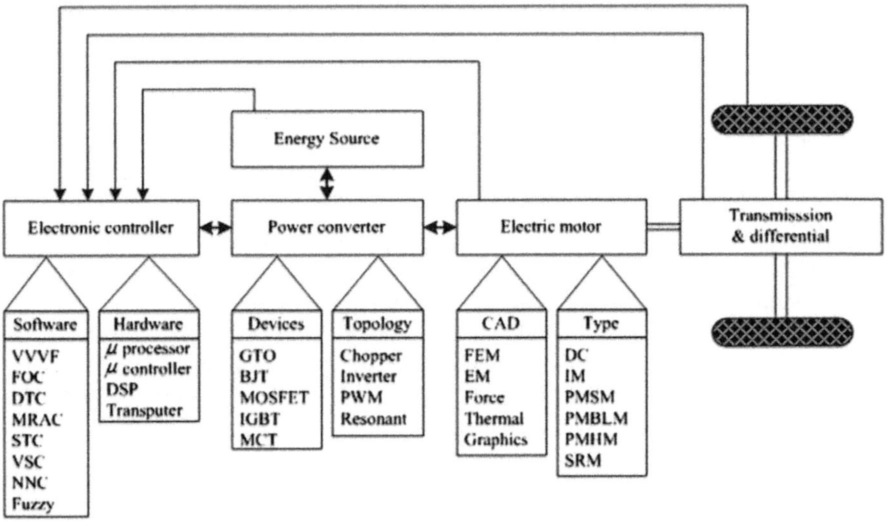

Fig. 2 Functional block diagram of an EV drive system

reliability and simplicity in both power converter disposition and motor structure, admirable thermal distribution, wide speed range and efficient regenerative braking, it's feasible for the Switched reluctance (SR) motors to provide superior characters for EV applications. Nevertheless, they may produce some dissatisfied problems such as torque ripples and acoustic noise.

Determined by three critical factors-vehicle constraint, driver expectation and energy source, the selection of traction motor drive is an important part in the EV manufacture. A driving profile including the climbing capability, braking ability, driving range, acceleration time and maximum speed is a critical definition for the drivers' anticipation. When considering the vehicle constraint, the vehicle weight, vehicle type and payload is the main concern. And when discussing the energy source, the fuel cells, flywheels, capacitors, batteries and various hybrid sources will be mentioned. Hence, it's necessary to operate at the system level when identifying the favored advantages and packaging choices for electric motor drive. To make sure the system operates functionally, the examination of the interplay between subsystems and those possibly impacts of system trade-offs is indispensable.

6 EV Near-Term Advanced Batteries

By recent batteries are meant state-of-the-art batteries that are likely to being targeted for vehicle applications in 5–10 years. The chemistry of these batteries is going to expand from the nowadays used lithium batteries and expected have energy densities <300 Wh/kg for cells.

Most of the current research is targeting to improve the energy density of lithium-ion batteries. Three aspects are studied in the research: (i) expand the specific charge range (mAh/g) of the cathode (positive electrode), (ii) boost the definite charge capacity of the anode (negative electrode), and (iii) develop the cell voltage. With respect to the application of vehicle, the performance of the cells should be increased while the high cycle life and sufficient power capability stay the same. The specific charge capacity of the electrodes relies mainly on the materials used and the cell voltage relies mainly on the electrolyte utilized. Therefore, the research is focused on the electrochemistry of electrode materials and organic electrolytes for reversible intercalation of lithium ions in those materials. The research targets are to search materials that have greatly higher mAh/g capacity than graphite currently used in the anode and the combinations of metal oxides used in the cathodes in presently available lithium batteries. The cell voltage relies on not only the characteristics of the electrolytes, but also their compatibility with the electrode materials. Nowadays, cell voltages for lithium batteries vary from 3 to 4 V. The current research is purposely designed cells with voltages not more than 5 V. The common method for each electrode is to prepare material composites, which obtain limited charge abilities (mAh/g) that are greatly larger than that in the current baseline. In the regard of the anode, the composites are composed of mixtures of graphite/carbon and silicones. Regarding to the cathodes, the composites incorporate layered atomic structures.

Various metal oxides—Mn, Ni, Co, and Al. Current lithium cells adopt materials that have charge capacity with value range varies between 100 and 150 mAh/g, and state-of-the-art materials having charge capacities not more than 250 mAh/g are being promoted. As for the anodes, most cells currently adopt graphites having limited charge capacities vary between 330 and 350 mAh/g. In conclusion, the most up-and-coming of the new anode materials are likely to be mixtures of carbon and silicone (Si/C). By far, the research suggests to reveal that mass fractions of silicone of 25–40% are feasible consistent with reasonable cycle life. As a result, the mAh/g of the anodes would be in the range of 800–1000. The LTO (Lithium Titanate Oxide) anode batteries have the features of quick charge, withstanding high and low temperature operating conditions and long cycle life. However, the cell voltage and energy density are lower.

7 Commercialization Roadmap

The major contributions to the success of the penetration of electric vehicles can be summarized into the following four categories: good product of high price-performance ratio; infrastructure of good economic efficiency; and innovative business model that can leverage the initial cost of batteries. The main standards that contribute to the success of commercialization of electric vehicles are cost, usage convenience, energy consumption, and emission level. For the sake of cost-efficiency, the residual value of batteries is one of the cores. In fact, batteries play an important role in not only supplying energy and power but also serving as energy storage, which can interact with the grid if it is not on the driving mode. Hence, it is the electric power industry that is responsible of exploiting the value of the batteries and optimizing the usage of batteries around the world. In order to achieve this target, the cooperation between the automotive industry and electric power industry is extremely necessary.

For the purpose of maximizing the added value of electric vehicles, it is of great significance to integrate electric vehicles with smart grid, and also with information and communication technology. Generally, the batteries of an EV can be used for the energy and power storage for EV. Additionally, it can be also used for the spinning reserve of power grid. For instance, the batteries can be used as energy sources when the grid is weak, on the contrary, the batteries can be charged when the grid is strong. Therefore, it has the following three main advantages, namely, higher stability and load factor of the grid, longer lifespan of the batteries, less loss of wind power and solar power.

In China, in the next ten years, the automobile production and sale will continue to grow with annual rate of about 10%. Therefore, the pressure to energy security and air pollution is more severe than other countries. Therefore, China will be the largest market of electric vehicles and has potential to be the kingdom of electric

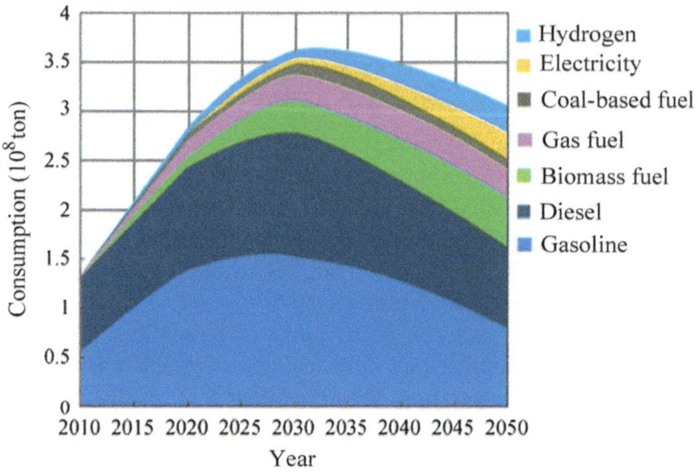

Fig. 3 Impacts of EV penetration on energy consumption

vehicles. The strategic development of electric vehicles in China can be divided into three phases. Phase 1 is from now to 2020, aiming to reduce air pollution and PM2.5, penetration rate will be 2%, mainly in public transportations, taxi, small electric vehicles. Phase 2 will be from 2020 to 2030, aiming to reduce total road transportation energy consumption by increasing fuel economy; penetration rate will reach 15%, mainly in urban area various applications. Phase 3 will be from 2030 to 2050, aiming to reduce CO_2 emission, penetration rate will reach 50%, and fuel cell electric vehicles will be commercially available. Figure 3 shows the impact of electric vehicle penetration on the energy consumption based on the above said three phases. The turning point will be in 2030, where the road transportation total energy consumption and air pollution will reduce, and the road transportation energy spectrum will move toward non fossil fuel.

8 Conclusion

Owing to the growing attentions on the environment protection and energy utilization, the development of EVs, HEVs and FEVs is increasing. The commercialization of the EVs and HEVs is no longer a dream. EVs and HEVs are ready for the domestic world. This article gives the reader the overview of the development of EVs, HEVs and FEVs with focus on engineering philosophy, technical features and commercialization roadmap. Upon the support of excellent product at reasonable cost, satisfactory infrastructure with economic, efficient and convenient performances and innovative business model, it is believed that the EVs market can be boosted up accordingly. It is very intriguing to see the rapid enhancement of EVs that will certainly improve the whole world in the near future.

References

1. C.C. Chan, “Renaissance scientists and engineers - Mass, Energy and Information”, Studies in Science and Technology, Volume 2, Number 2, 2013.
2. C.C. Chan and K.T. Chau, Modern Electric Vehicle Technology. Oxford UK: Oxford University Press, 2001.
3. Z.Q. Zhu and D. Howe, “Electrical Machines and Drives for Electric, Hybrid, and Fuel Cell Vehicles,” Proc. IEEE 95: 746–765, 2007.
4. Andrew Burke, “Advanced Batteries for Vehicle Applications”, Encyclopedia of Automotive Engineering, Hybrid and Electric Powertrain, Energy Source Batteries. Wiley, 2015.

Smart Mobility: Driver State Estimation and Advanced Driver-Vehicle Interfaces

Lucas P.J.J. Noldus, Andrew J. Spink, Ramon Bollen and Tobias Heffelaar

Abstract With increasingly complex user interfaces and advancing automation, measuring the state of the driver has never been more important. By using sensor fusion techniques we combine information from multiple sources to accurately and robustly measure the driver's state—drowsiness and attention, workload and cognitive load, pleasure and anxiety. Only by integrating and synchronizing all the data streams can we properly understand the interaction of the driver and the vehicle.

Keywords Driver · Behavior · Human factors · Simulator · Eye-tracking · Multi-modal data integration

1 Introduction

With the trend towards automated vehicles, the role of the driver is changing. This means that all sorts of new questions arise with respect to studying driver-vehicle interaction. How can the driver keep alert enough to intervene at the moment he or she needs to override the automatic system? What is the experience of the driver and passengers in an advanced car—do they feel safe, trusting the automation, or helpless and nervous? Measuring the state of the driver has never been more important, and this has led to a demand for new methods and tools. At the same time, it has become clear that to fully understand the driver's state, it is not sufficient to rely on a single measure, but use of sensor fusion techniques combining multiple modalities is necessary. In the Netherlands, several public-private partnerships have been formed between knowledge institutes and corporations to address these challenges in a wide variety of settings, from driving simulators to instrumented cars, to fully autonomous self-driving vehicles on the public road. In those cases we and our partners have developed tools to measure the driver's state by combining state-of-the art measurements including eye tracking, physiology

L.P.J.J. Noldus (✉) · A.J. Spink · R. Bollen · T. Heffelaar
Noldus Information Technology BV, Wageningen, The Netherlands
e-mail: L.Noldus@Noldus.nl

M.C. Tandon and P. Ghosh (eds.), *Mobility Engineering*,
DOI 10.1007/978-981-10-3099-4_2

(heart rate, skin conductance), facial expression and behavior of the vehicle/simulator. Without integration and synchronization, these data are much less useful. For instance, if a driver looks at a point on the display of a simulator and the simulator displays a person at that point, it is only by combining the data that you can know that that the driver has seen the pedestrian. These techniques have proved so promising that they have been incorporated into commercial products currently being released onto the market. Besides driver studies, these techniques are applied to a variety of other domains, such as consumer behavior in virtual supermarkets.

2 Why Measure Driver Behavior in a Simulated Environment?

Measuring driver behavior in a virtual (simulated) environment has a number of advantages over similar measurements in an actual car:

- It is possible to run exactly the same scenario for multiple subjects (for example precisely the same traffic on the road).
- You can include scenarios which would otherwise be too dangerous (for example a pedestrian stepping out in front of the car).
- It is possible to test scenarios which normally only rarely occur (such as failure of a component of the car).
- You can more cheaply and easily test scenarios for which you want to test several variants (e.g. several UI designs).
- It can also be easier to instrument a simulator with measurement devices such as physiological sensors and eye trackers than a real car.

However, if data from those sensors are not integrated and synchronized with the simulator data, then interpretation is limited and difficult. In this paper we present an integrated system, named DriveLab, which has been designed to overcome precisely that issue.

3 DriveLab System Description

DriveLab [1, 2] has been developed in the ADVICE project (see Acknowledgments). It is built with a modular architecture around an integration and synchronization platform. There are a number of core components, fundamental to the system, and a variety of optional components which have been tested for compatibility with DriveLab, but are not required for its operation. It is an open system, so that other modules can be added as required (Fig. 1).

Fig. 1 DriveLab in operation

3.1 *Core Components*

3.1.1 Driving Simulator

A core component of a DriveLab setup is the driving simulator. The first DriveLab implementation is based on a Green Dino simulator. Green Dino BV is a Dutch developer of driving simulators for both instructional and research purposes. The driver is seated with a physical steering wheel and controls such as pedals and hand brake, and both the instrument display of the car and the external view (windscreen, side windows and mirrors) are created virtually. The simulator is programmed with a number of predefined routes and events, which can be modified by the user, and all data generated by the simulator is available to the rest of the DriveLab system.

3.1.2 Eye Tracker

The Swedish company Smart Eye AB produces high-end 3D remote eye trackers. These have a number of features which make them especially suitable for automotive applications. They can use multiple cameras, so that a wide field of view (up to 360°) can be encompassed, giving the test subject complete freedom of head rotation. In the DriveLab system we use an eye tracker with three cameras. 3D depth is correctly interpreted (e.g. the system knows if the driver is looking at the dashboard or out of the car and there is no parallax effect). The eye tracker gives data output for both the head, left and right eye with over 145 parameter values including gaze tracking, head tracking (6DOF), eyelid tracking, pupil tracking, raw and filtered gaze, blinks, fixations, and saccades.

3.1.3 Audio-Video Recording and Behavioral Annotation

The driver's body posture and gestures are captured with a scene camera, mounted behind the driver and looking forward over his shoulders. The video stream is recorded with Noldus' Media Recorder software, and this is fed into Noldus' The Observer XT software, where it can (optionally) be annotated by the researcher to log specific behavioral events. The scene video is shown with a gaze overlay superimposed from the eye tracking data.

3.1.4 Integration and Synchronization

A key development in the ADVICE project was a communication protocol between all the various components, named N-Linx. N-Linx enables data from all the different components to be streamed real-time. It uses the RabbitMQ messaging framework as a platform for sending and receiving messages. RabbitMQ was selected as it is open source software supporting all major operating systems and most development environments (see Fig. 2). It is also scalable (more servers can be added) and can deal with lost connections without losing data.

An example of the use of N-Linx is the automated tracking of the driver's visual fixation on moving objects. The driver's point of gaze coordinates, recorded by the eye tracker, are sent (using N-Linx) to the simulator, which returns the identity of the virtual object being displayed at that specific position. Thus, the system can track which objects (UI items, road signs, cyclists, etc.) the driver looks at. The combined data is then sent back via N-Linx and then, together with the rest of the

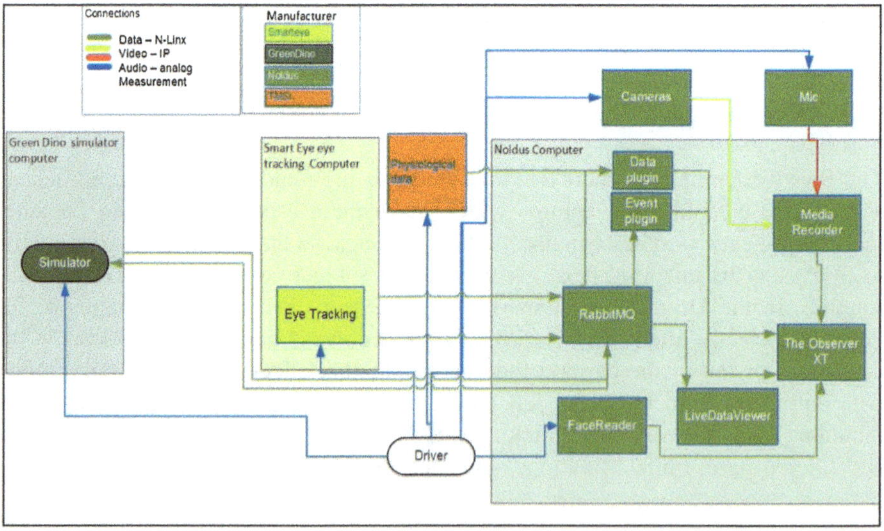

Fig. 2 Schematic diagram of the communication and architecture of DriveLab

data streams (facial expressions, physiological data, etc.), sent to The Observer XT. The Observer XT is Noldus' software for behavioral coding, which also serves as a data integration platform [3].

3.1.5 Data Analysis

In The Observer XT all the data can be analyzed. In particular, a selection can be made of particular persons, behaviors, and time slices relevant to study; for instance the eye blink rate whilst carrying out a particular task or the speed of the car when the simulator presented a road sign that was not observed (for instance if the driver was looking at the dashboard).

3.2 Optional Components

3.2.1 Physiology

Measures such as driver alertness, workload and stress can most effectively be measured by combining physiological data with other measures (such as pupil diameter and blink rate). DriveLab uses The Observer XT to integrate and synchronize the physiological data with the other data streams. This means that the system can work with virtually all commercial physiological data acquisition systems. In the ADVICE project, a TMSi Porti (www.tmsi.com) was used. A similar setup has also been tested with Near Infra-Red Spectroscopy (NIRS) equipment manufactured by Artinis Medical Systems (www.artinis.com). Both TMSi and Artinis support the N-Linx protocol.

3.2.2 Emotion

The emotional state of the driver can be measured by video analysis of their facial expressions with Noldus' FaceReader software. This software measures the six basic emotions (happiness, sadness, surprise, fear, disgust and anger) as well as neutral [4]. In addition it gives a measure of both the valance (balance of positive and negative emotions) and arousal as well as other measures such as the opening or closing of eyes or mouth, raising of eye brows, and facial action units [5] (which indicate more subtle emotions, which could be relevant in this context).

3.2.3 Live Data Viewer

Noldus also developed special software (named Live Data Viewer) for customizable live visualization of all data streams (Fig. 3). This has a two purposes. Firstly,

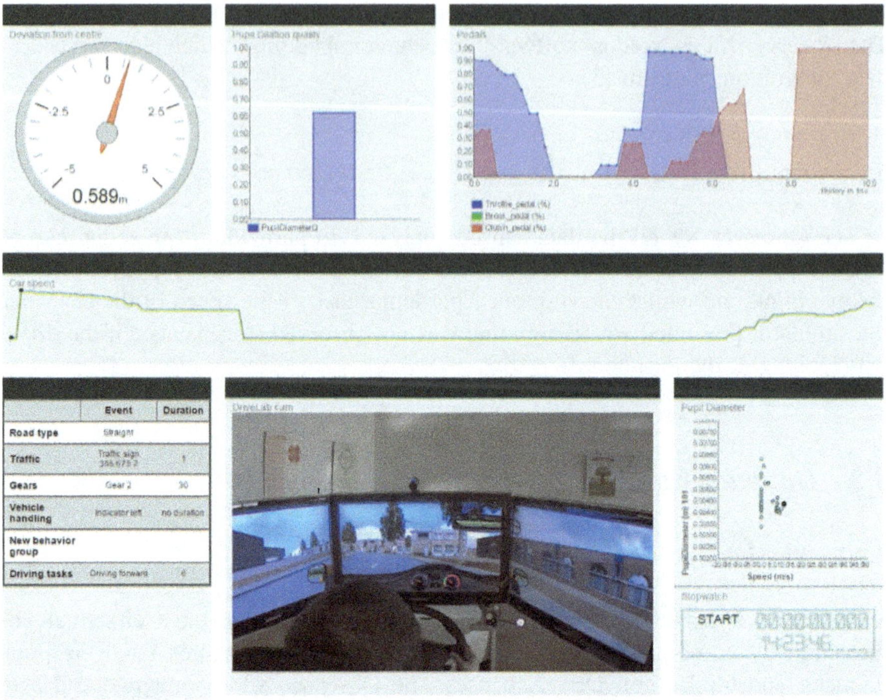

Fig. 3 DriveLab live data viewer showing data from three sources (simulator, eye tracker and video camera) with a variety of different visualizations

to enable the researchers to monitor the measurements, for instance to check that all the instrumentation is working correctly. Secondly so that the test leader can react appropriately to specific events in real-time, for example to give feedback to a trainee or to alter the experimental conditions accordingly. The Live Data Viewer can be configured so that the user can define which data sources he or she wants to see, and also which visualization of the data is most relevant, as well as setting a number of parameters such as the maximum and minimum data values. The Live Data Viewer runs in a browser on PCs, notebooks, tablets or smartphones, allowing researchers to monitor DriveLab sessions in the lab or from a remote location.

4 Conclusions and Future Developments

The methods presented in this paper demonstrate the possibility and the added value of measurements which do not just measure one aspect of a person's behavior, but integrate several different channels (sensor fusion) to give a more robust and more complete behavior. The upcoming Internet of Things and associated increasing availability of relatively cheap and abundant sensors gives many possibilities in this respect. The partners in ADVICE are also working in other projects on applying the same technology to different domains, for instance we are studying consumer behavior in a virtual shop [6] with the addition of EEG and NIRS [7] measurements.

Within the project the main focus now is to further develop it so that DriveLab also works in an actual car as well as a simulator [8]. Although simulators have a number of advantages, there are also disadvantages such as simulator sickness [9] and validity [10]. We have also shown the validity of DriveLab or its components in several practical use-cases [11].

Acknowledgments DriveLab has been developed in the project ADVICE (Advanced Driver Vehicle Interface in a Complex Environment) [12], which is funded by the Dutch Ministry of Economic Affairs and will finish in December 2015. The partners in the project are: HAN University of Applied Sciences (www.han.nl/international/english/), Delft University of Technology (www.tudelft.nl/en/), Noldus Information Technology (www.noldus.com), TNO (www.tno.nl/en/), and TomTom (www.tomtom.com).

References

1. DriveLab website (2015). Retrieved 16 June 2015 from http://www.noldus.com/innovationworks/products/drivelab.
2. Tobias Heffelaar, Jorrit Kuipers, Jonas Andersson, Leon Wiertz & Lucas Noldus (2014). Easy to Use Driving Behavior Analysis Using DriveLab. HCI International 2014 - Posters' Extended Abstracts. Springer International Publishing. Pp 330–334.
3. Patrick H. Zimmerman, Liesbeth Bolhuis, Albert Willemsen, Erik S. Meyer, and Lucas P.J. J. Noldus (2010). The Observer XT: a tool for the integration and synchronization of multimodal signals. Proceedings of Measuring Behavior 2008 (Maastricht, The Netherlands, August 26–29, 2008). Eds. A.J. Spink, M.R. Ballintijn, N.D. Bogers, F. Grieco, L.W.S. Loijens, L.P.J.J. Noldus, G. Smit, and P.H. Zimmerman. Pp 125-126.
4. Lewinski, P.; den Uyl, T.M.; Butler, C. (2014). Automated facial coding: validation of basic emotions and facs aus in FaceReader. *Journal of Neuroscience, Psychology and Economics*, 7 (4), 227–236.
5. P. Ekman and W. Friesen. *Facial Action Coding System: A Technique for the Measurement of Facial Movement.* Consulting Psychologists Press, Palo Alto, 1978.
6. FOCOM website. Retrieved 16 June 2015 from: http://www.noldus.com/projects/focom-food-and-cognition-model-systems.
7. Villringer, J. Planck, C. Hock, L. Schleinkofer, U. Dirnagl (1993), Near infrared spectroscopy (NIRS): A new tool to study hemodynamic changes during activation of brain function in human adults. *Neuroscience Letters*, 154 (1–2), 101–104.
8. Tobias Heffelaar, Renske B. Landman, Menno Merts, Jasper Michiel van Hemert, Arjan Stuiver, Lucas Noldus (2014). Driver state estimation: from simulation to the real world. Proceedings of Measuring Behavior 2014, (Wageningen, The Netherlands, August 27–29, 2014). Eds: A. Spink, L. Loijens, M. Woloszynowska-Fraser & L Noldus.
9. Brooks, J.O., Goodenough, R.R., Crisler, M.C., Klein, N.D., Alley, R.L., Koon, B.L., & Wills, R.F. (2010). Simulator sickness during driving simulation studies. *Accident Analysis & Prevention*, 42(3), 788–796.

10. Bédard, M.B., Parkkari, M., Weaver, B., Riendeau, J., Dahlquist, M. (2010). Assessment of driving performance using a simulator protocol: Validity and reproducibility. *American Journal of Occupational Therapy,* 64, 336–340.
11. Kuipers, J. (2014). Automation of training and assessment with DATA Centred Design (DCD). Proceedings of Measuring Behavior 2014, (Wageningen, The Netherlands, August 27–29, 2014). Eds: A. Spink, L. Loijens, M. Woloszynowska-Fraser & L Noldus.
12. ADVICE project (2015). Retrieved on 16 June 2015 from http://www.noldus.com/projects/advice.
13. Kuipers, J. (2014). Multi variable strategy reduces symptoms of simulator sickness. Proceedings of Measuring Behavior 2014, (Wageningen, The Netherlands, August 27–29, 2014). Eds: A. Spink, L. Loijens, M. Woloszynowska-Fraser & L Noldus.

Modular Hydraulic Trailer Technology

Jochen Landes

Abstract Transportation of overweight and over dimensional (OWC/ODC) consignments with more than several hundred tons is a big challenge for heavy duty freight haulers. Sturdy and reliable modular trailers from TII Group ensure an exceptional logistics performance. Transportation technology with hydraulic axles is the key to protect the Indian infrastructure with its public roads, bridges, places, etc.

Keywords Goods transport · OWC/ODC · Hydraulic axles · Hydraulic support groups · Engine supported trailers

1 Introduction

Transportation of overweight and over dimensional (OWC/ODC) consignments with more than several hundred tons is a big challenge for heavy duty freight haulers. Sharp curves, steep inclines, frequented motorways, bridges, subways and narrow streets have to be negotiated and adjustments have to be made to suit rough terrain conditions. With advanced and reliable products from TII Group, Germany, exceptional logistics performances can be attained due to their extreme reliability and, thus, the risks of long-distance transportation can be considerably reduced.

The hydraulically supported pendulum axle was designed by TII Group's SCHEUERLE in 1956. It has revolutionized heavy duty transportation around the globe and today it is the basis for heavy duty road traffic solutions worldwide. It allows for an even distribution of the payload on several axles by forming so-called support groups. In order to keep the payload within the technical and approved specific maximum load bearing capacity limits of axles, lightweight design while retaining the rigid main beams is very important.

J. Landes (✉)
TII India Pvt Ltd., Bawal, India
e-mail: jochen.landes@tii-india.com

M.C. Tandon and P. Ghosh (eds.), *Mobility Engineering*,
DOI 10.1007/978-981-10-3099-4_3

Fig. 1 Typical transportation task

Figure 1 shows a typical task for a heavy duty logistics expert, a 300–500 t transformer moved across a fragile bridge. To avoid possible human fatalities in this case, the trailers are equipped with a Power Pack Unit for self-propelled operation.

2 Hydraulic Axles

The hydraulically supported pendulum axle is the basis for all on- and off-road solutions from TII Group. It consists of a bogey with upper and lower and arm and a suspension cylinder which is connected to a hydraulic system. Figure 2 shows the functional principle. Under the conditions shown, the hydraulic fluid is pushed out of the cylinder of one bogie and flows into the cylinder of the other bogie. Thus, both axles keep ground contact and the load per axle remains constant. Without pendulum axles with hydraulic suspension, some wheels may lose contact and cause overloading of the remaining wheels. Overloading of wheels in turn results in

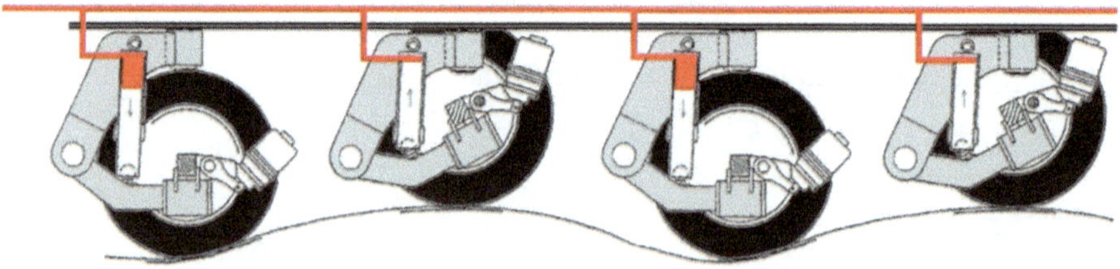

Fig. 2 Hydraulic pendulum axles

Fig. 3 Hydraulic pendulum axles in rough terrain

eventual road surface damage. Such overloading is fatal in case of bridges. Figure 3 shows the behaviour under rough terrain conditions. Under almost all circumstances all wheels will remain in ground contact.

3 Hydraulic Support Groups

To improve stability the trailer bogies can be hydraulically combined in so-called support groups, as shown in Fig. 4. Depending on the load configuration, either a 4-point or 3-point configuration can be selected. Like a chair on legs, the load can be levelled in the best possible way and lifting and lowering can be done very precisely. A 4-point support group shows the best possible protection against tilting. However, torsion along the trailer may occur. Thus, in some cases, a three point circuit is a better choice. It has less stability but avoids the torsion problem.

In general, 4-point configuration will be selected for loads with high centre of gravity or with banking (side slope) of the road. 3-point is the choice for payloads with low centre of gravity or for loads which need to be protected from torsion of the trailer.

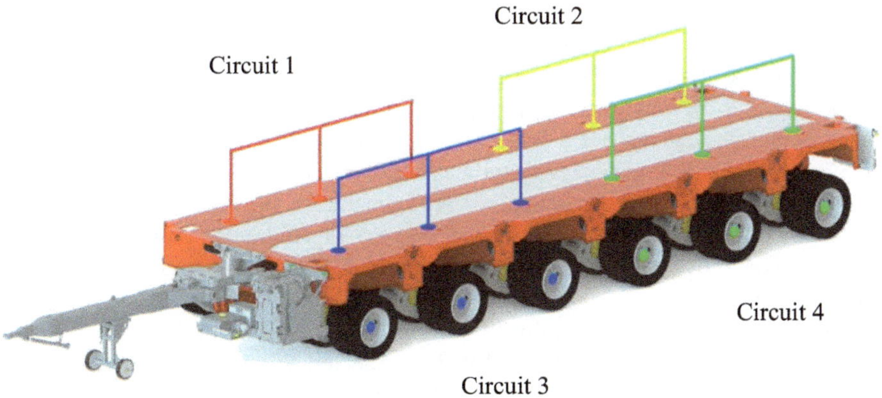

Fig. 4 Hydraulic support groups

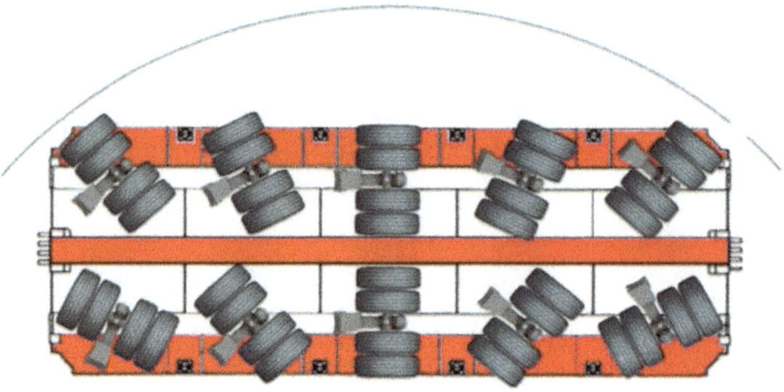

Fig. 5 Steering

4 Steering

To ensure safe transport of heavy loads even under severe conditions like steep inclines, rough terrain or narrow bends in small villages, a sophisticated steering system is required. In case of hydraulic trailers from TII Group the maximum steering angle is up to 60°. The steering can be changed and adjusted easily from the top, depending on the requirement from prime mover and transportation task. The various wheels run on circles with different radius to ensure maximum flexibility, as shown in Fig. 5.

5 Engine Supported Trailers

The basic version of a hydraulic trailer is a passive vehicle, towed by a prime-mover. The hydraulic systems such as steering and suspension can be regulated when needed by an external Hydraulic Power Unit. Some of the prime movers have such unit. However, for certain applications, where prime mover doesn't have a Hydraulic Power Unit or has only a weak power unit, a separate engine needs to be attached to the trailer. For best results and depending on the transportation task, there are a few basic options possible.

A small engine/pump unit with typically around 10–22 kW, called Power Pack Unit (PPU) can be mounted on the trailer. It generates sufficient hydraulic pressure to lift, lower and steer the trailer in the absence of Hydraulic Power Unit of the prime mover. However, self-propelled operation is not possible with this configuration.

Self-propelled operation requires a more powerful engine. In this case, a Power Pack Unit in the range of 150–200 kW is connected to enable hydrostatic drives mounted in the wheels (Fig. 6, see wheels in the middle). These consist of a series of hydraulic pistons in a 'star-array', operated by pressure generated in the PPU. In

Fig. 6 Hydraulic trailer with power booster

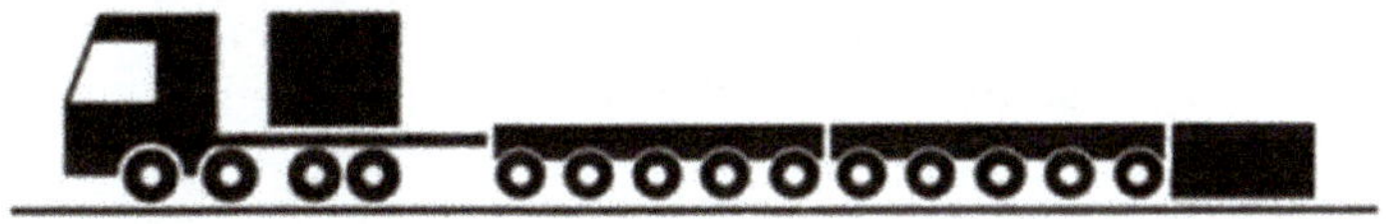

Fig. 7 Power booster in highway mode (PPU 'OFF')

normal highway mode the PPU is switched OFF, the drives are in idle position and no additional driving power is generated. There is no influence on the trailer operation and maximum speed of 80 km/h can be reached (Fig. 7).

With PPU 'ON' the Power Pack Unit can be used as traction support for steep inclines. Instead of adding more prime movers, PPU enables the hydrostatic drives in the wheels and generates sufficient power to support the main prime mover engine to manoeuvre (Fig. 8).

While transporting over fragile bridges, PPU is used instead of a prime mover in order to reduce the length and gross weight of the train. The PPU can be switched 'ON' and propels the trailer forward at a low speed of max 12 km/h. This mode is also very useful when smooth movement conditions are required (no sudden acceleration from prime mover). Figure 9 shows a typical configuration.

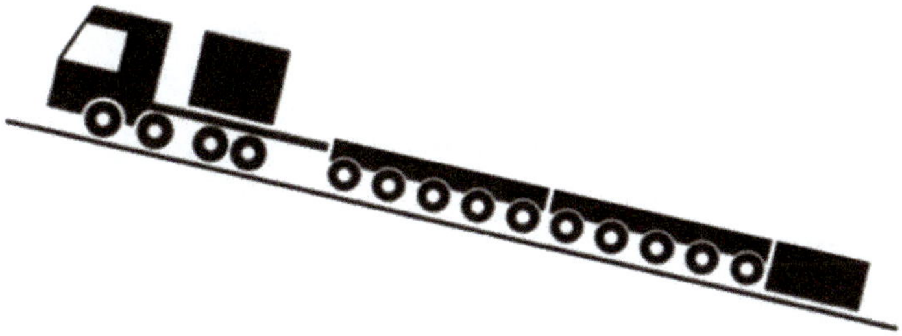

Fig. 8 Power booster in incline mode (PPU 'ON')

Fig. 9 Power booster in self-propelled mode (PPU 'ON')

6 Other Applications

Wind power is becoming increasingly important. TII offers special trailers for rotor blades, wind tower segments, and other bulky components—for on shore and for off shore. Sophisticated wind blade adapters for operation in areas with steep inclines, sharp curves and narrow bends further help transporters to simplify their operation. The adapter swings the blade upright, so that the overall length is drastically reduced from e.g. more than 50 m to around 10 m. After passing the critical area, the blade can be turned back into horizontal position.

7 Conclusions

Transportation of overweight and over dimensional (OWC/ODC) consignments with more than several hundred tons is a big challenge for heavy duty freight haulers. Sharp curves, steep inclines, frequented motorways, bridges, subways and narrow streets have to be negotiated and adjustments have to be made to suit rough terrain conditions.

Since its market launch, the hydraulically supported pendulum axle, which was designed by TII Group's SCHEUERLE in 1956, has revolutionized heavy duty transportation worldwide. Sophisticated technology like advanced steering systems and detachable Power Pack Units for self-propelled operation allow logistics experts to transport their goods safely over long distances and without any hassles.

References

1. K25H – An innovative vehicle concept from SCHEUERLE and KAMAG, TII Group, Germany
2. SPMT – Self-Propelled Modular Transporters, TII Group, Germany
3. TII Group – Global leader in heavy duty transportation accords top priority for Indian market, Motor India, April 2014

Current Situation and Prospect of Transportation Development in China

Lu Huapu

Abstract Since reform and opening-up policies are taken in China from 1978, dramatic changes happened in transportation field. The present situation of transportation is introduced in general firstly. Then, the key new technologies achieved are summarized briefly. Finally, this paper briefly describes the development strategy and policy prospect of China's transportation.

Keywords Transportation · Urban traffic · High speed railway · Transportation policy

1 Introduction

Since the reform and opening-up, the transportation industry in China meets a huge chance of development, especially in the field of railway, highway, port and aviation. In recent years, with the diversification of people's travel requirements, fields such as intelligent traffic management have also developed rapidly. In the face of future traffic development, advice as well as related policies should also be proposed to meet future traffic requirement.

2 Current Situation and Development Strategy

2.1 Current Situation of Urban Cities and Transportation Development in China

By the end of 2014, the number of China cities is 653, urban population is 749 million and urban built-up area is 46,000 km^2, which is an increase of 3.4 times,

L. Huapu (✉)
Director of Institute of Transportation Engineering, Tsinghua University, Beijing, People's Republic of China
e-mail: luhp@mail.tsinghua.edu.cn

M.C. Tandon and P. Ghosh (eds.), *Mobility Engineering*,
DOI 10.1007/978-981-10-3099-4_4

4.3 times and 9.1 times in comparison with the number in 1978 respectively, when the reform and opening-up just begins (by the end of the year 1978, the number of China cities was 194, urban population was 172 million and urban built-up area was 5000 km^2).

By the end of May, 2015, the national motor vehicle ownership is 270 million, in which car ownership is 162 million, making 59.91% of motor vehicle ownership; small passenger car ownership is 125 million; motorcycle ownership is 90.64 million, making 33.59% of the national motor vehicle ownership. In comparison with the year 1978, the motor vehicle ownership of 2014 is 199 times of that in 1978, small passenger car ownership is 483 times of that in 1978. We can see it clearly in Fig. 1.

2.2 *Railway*

By the end of 2014, China railway mileage reaches 112,000 km and electrification mileage reaches 65,000 km, accounting for 58.3% of the railway mileage, and the electrification mileage meets an increase of 16.9% in comparison with last year. Newly constructed railway mileage reaches 8,427 km and the high speed railway operating mileage is 16,000 km (accounting for more than 60% of the world. Railway construction investment reaches 808,800,000 Yuan.

National railway locomotive ownership is 21,100 units, in which 8423 units are high power locomotive (CHINA RAILWAY HIGH-SPEED) named as "harmony". Railway passenger vehicle ownership is 60,600 units and railway wagon vehicle ownership is 710,100 units (Fig. 2).

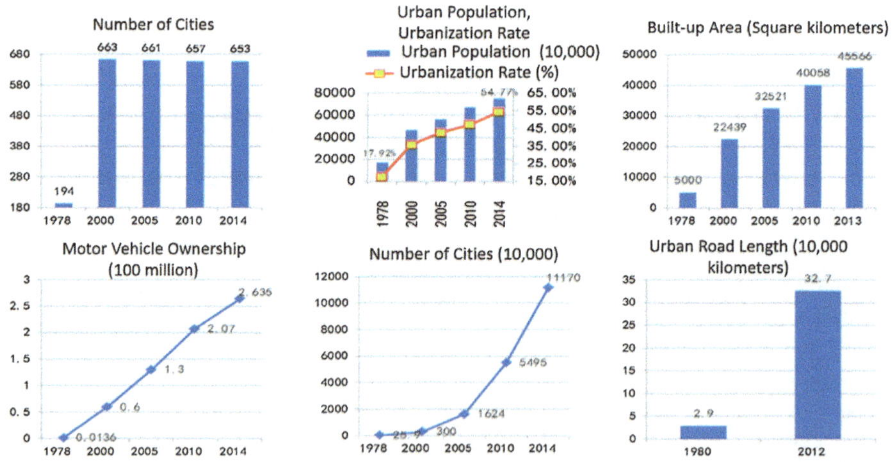

Fig. 1 Huge change of China's cities since the reform and opening-up

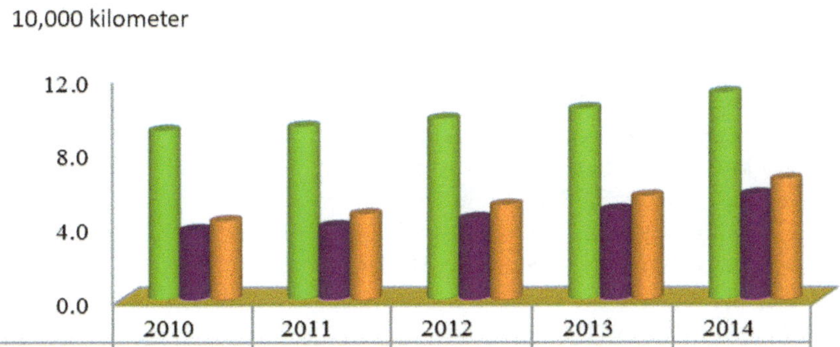

	2010	2011	2012	2013	2014
Operation Mileage	9.1	9.3	9.8	10.3	11.2
Double Track Mileage	3.7	3.9	4.4	4.8	5.7
Electrification Mileage	4.2	4.6	5.1	5.6	6.5

Fig. 2 China railway operation mileage status

In the year 2014, 55 railway technical standards containing ***High Speed Railway Design Code, Inter City Railway Design Specification*** were published, English translation and publication work for 54 technical standards were completed, ten international standard setting works of International Union of Railways (UIC) and nine that of International Electro technical Commission(IEC) were also hosted by China.

In 2014, five scientific and technological achievements in railway industry won the 2014 National Science and Technology Award, which are ***Research and Application of a Series of High Power AC Drive Electric Locomotive Based on Autonomous Technology Platform, Comprehensive Inspection and Assessment of Key Infrastructure Facilities in High Speed Railway, Complete Engineering Technology for Super-Large Diameter Yangtze River Shield Tunnel on High Water Pressure, Shallow Overburden Soil Complex Terrain Geology, Upgrade Version Project of China Railway Technology Innovation System.*** Four patents were awarded as the Sixteenth China Excellent Patent Award.

China's Qinghai-Tibet Railway overcomes three major world engineering problems: permafrost, alpine hypoxia, ecological vulnerability, which is the unprecedented great feats in the history of human railway construction. High speed railway's independent environmental protection technology and the role of improving local economic and social development have been recognized internationally. Beijing-Shanghai high-speed railway project adopts independent innovation of high speed railway technology including soft soil foundation reinforcement, rail technology, communication system, signal system operation, control mode and etc. Beijing-Shanghai high-speed railway project eases the tension of Beijing-Shanghai channel capacity, gains remarkable social and economic benefits and reaches world-class level in independent innovation technology.

2.3 Highway

By the end of 2014, the national highway mileage reaches 4,463,900 km, the national highway density is 46.50 km/100 km^2, and expressway mileage reaches 111,900 km. National rural highway (including county road, township road, and village road) reaches 3,881,600 km (Fig. 3).

China's highway construction technology has reached world advanced level, the following are some cases: Su Tong Bridge, located in Nantong City of Jiangsu Province and Suzhou (Changshu) city, with a total length of 32.4 km, is the largest bridge construction project and is built in the most complex construction conditions in the history of China bridge construction. It holds four world records including "The maximum group pile foundation", "Tallest bridge towers (300.4 m)", "Longest stay cable (557 m)", "Maximum main span".

Nanjing Yangtze River Bridge owns a total length of about 15.6 km, in which 4.7 km are crossing River Bridge. It is the first steel tower cable-stayed bridge in China and the first arc shaped steel tower cable-stayed bridge in the world.

Qinling-Mountain Zhongnan tunnel with a length of 18.02 km is the longest tunnel in Asia, ranking second after the longest tunnel—Norway Nordahl tunnel (24.51 km). It overcomes construction problems as fault, water inrush, rock burst and major operation technical problems as ventilation, fire and monitoring, and makes China's tunnel construction technology reaching a new level.

2.4 Port

By the end of 2014, the national inland waterway navigation mileage reaches 126,300 km. Mileage of waterway with grade is 65,400 km, accounting 51.8% of

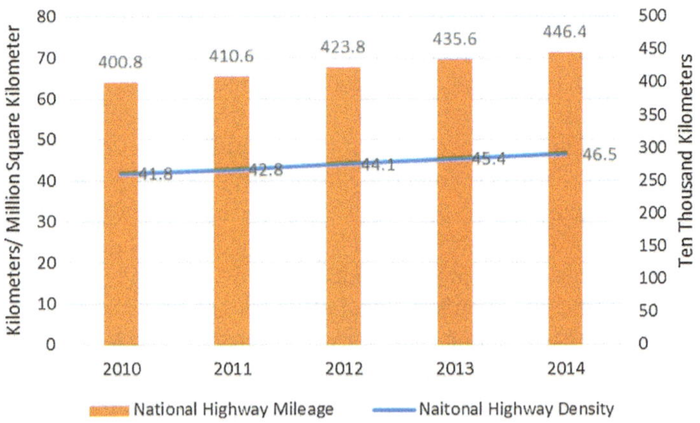

Fig. 3 China highway historical context

Fig. 4 Channel of each grade and their proportion

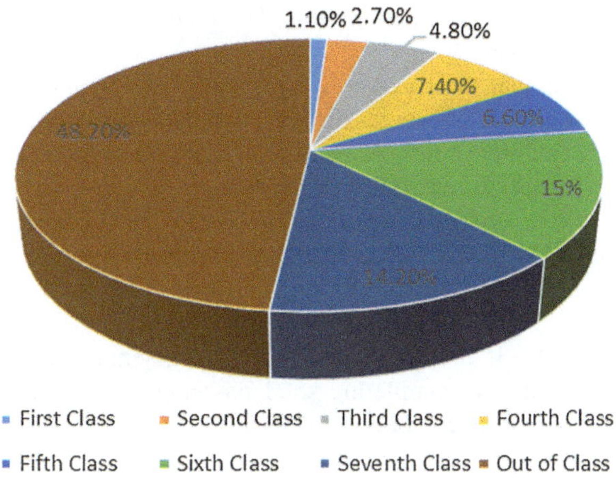

all waterway. The navigable mileage of inland waterways in each grade are: First grade channel 1341 km, second grade channel 3443 km, third grade channel 6069 km, fourth grade channel 9301 km, fifth grade channel 8298 km, sixth grade channel 18,997 km, seventh grade channel 17,913 km. Substandard channel 60,900 km (Fig. 4).

By the end of 2014, the national port has 31,705 berths for production, among which, 5834 are costal ports for production and 25,871 are inland ports for production.

There are 2110 berths owning cargo handling capacity over one thousand tons national wide, among which 1704 are costal ports and 406 are inland ports.

Ports national wide finish cargo throughput of 12.452 billion tons in 2014, in which, costal ports complete 8.033 billion tons and inland ports complete 4.419 billion tons.

Port national wide complete the passenger throughput of 183,000,000 people, in which, costal ports complete 81,000,000 and inland ports complete 102,000,000.

According to 2014 World Port Cargo throughput ranking, China owns 8 of top 10 and Zhoushan port ranks first.

According to container throughput ranking, China owns 7 of top 10 and Shanghai port ranks first.

2.5 *Aviation*

By the end of 2014, there are 202 civil airports with certification including 200 airports and 198 cities with scheduled flight.

64 navigable airports have passenger throughput over one million, which is an increase of four in comparison with last year, 24 navigable airports have passenger

throughput over ten million; 50 navigable airports have cargo throughput over 10,000 tons.

National civil aviation completes passenger transport volume of 390,000,000 person-time, passenger turnover volume of 633.42 billion person*kilometer, cargo and mail transport volume of 5,941,000 tons and cargo and mail turnover volume of 18.78 billion tons*kilometer.

Civil aviation completes passenger throughput of 832 million person-time, which is an increase of 10.2% compared with last year; it also completes total cargo throughput of 13,561,000 tons, which is an increase of 7.8% compared with last year.

In 2014, Beijing Capital Airport passenger throughput ranked second for five consecutives, completing total passenger throughput of 86,128,000 and international passenger volume reached 20,730,000 person-time.

3 Current Situation of Urban Traffic Development in China

3.1 *Road Infrastructure*

From 2000 to 2014, urban road length increases from 29,000 km in 1980 to 32,700 km in 2014. Cities national-wide have applied a total number of more than 600,000 km for marking and added more than 700,000 traffic signs. Traffic channelization rate has reached 95.5%.

3.2 *Public Transportation Development*

3.2.1 Basic Situation

The urban passenger transportation system transports 131.566 billion passengers in 2014, which is an increase of 2.5% compared with last year. Buses and trolleybuses account for 59.4% of all passenger volume, which can be seen from charts (Fig. 5).

By the end of 2014, there are 22 cities in China which have built or operated rail transit system, with a total mileage of 3173 km. 40 cities' rail transit system are under construction with a total mileage of 4,073 km. In which, the mileage of inner-city rapid rail transit is 308 km, the mileage of Maglev is 30 km, the tram construction mileage is 141 km (Fig. 6).

By the end of 2014, China has 528,800 buses and trolleybuses, which is an increase of 3.8% compared with last year. Among all the buses and trolleybuses,

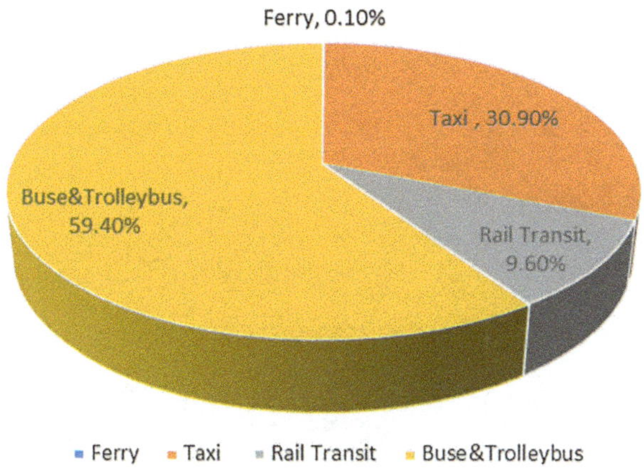

Fig. 5 Passenger volume components of urban transportation system

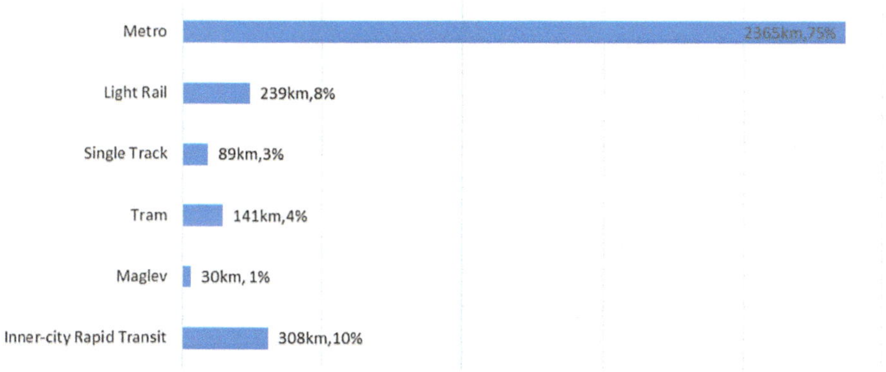

Fig. 6 Current situation of China cities' rail transit construction

3% use gasoline, 35% use natural gas and for the most, 62%, use diesel as power source.

There are currently 45,052 lines in operation for buses and trolleybuses national wide, with a total mileage of 817,800 km. The total bus station area is 52,310,000 m^2, in which, parking lot area is 43,750,000 m^2. China owns 6897.3 km of dedicated bus lanes, which is an increase of 1006.7 km compared to last year.

There are 27 Chinese cities that have already constructed and operated bus rapid transit (BRT), and the distribution map can be seen in Fig. 7.

Fig. 7 Chinese cities have constructed or operated BRT

3.3 Intelligent Transportation Management

By the end of 2010, 64% cities in China (423 cities) have achieved traffic signal network control, 82% cities have established traffic command center. In 2010, national intersection light control rate reaches 43.1%, and at the end of 2013 it reaches 91.2%, which is an increase of 110% compared to last year.

In comparison with 2005, 2010's number of command and control platform based on GIS increases from 35 to 200, and that of video surveillance point increases from 9250 to 51,000. The number of fixed collecting points for traffic flow information increases from 16,000 to 73,000 and satellite positioning device increases from 2057 to 29,000.

3.4 Walking and Bicycle

In 2012, the Development and Reform Commission put forward the 12th five-year energy saving emission reduction implementation plan. In this plan, it advocates "135" travel plan, which means choosing walking within 1 km travel distance, choosing bicycling within 3 km travel distance and choosing taking public transit within 5 km travel distance.

In the same year, Ministry of Housing and Urban-Rural Development, Ministry of Finance formulated the Guiding Opinions on strengthening the construction of urban pedestrian and bicycle transportation system. In this guiding opinion, it puts forward on proposing big cities, mega cities to develop walking and bicycle traffic, focusing on solving the middle and short distance travel and transfer to public connection. For small and medium cities, they should focus on the development of walking and bicycle traffic, which can be the main means of transportation. The guidance requests that by 2015, the city walking and bicycle travel environment should be significantly improved and bicycle sharing rate should be gradually increased. Different requirements were proposed for different size of cities. For urban cities with population more than 10,000,000, walking and bicycle travel share rate should be 45% or more.

To ease urban traffic congestion and solve "last-mile problem" for public transportation, 121 cities in China implemented public bicycle rental service project.

The number of urban bicycle rental points for top 10 cities in China are listed in Figs. 8 and 9.

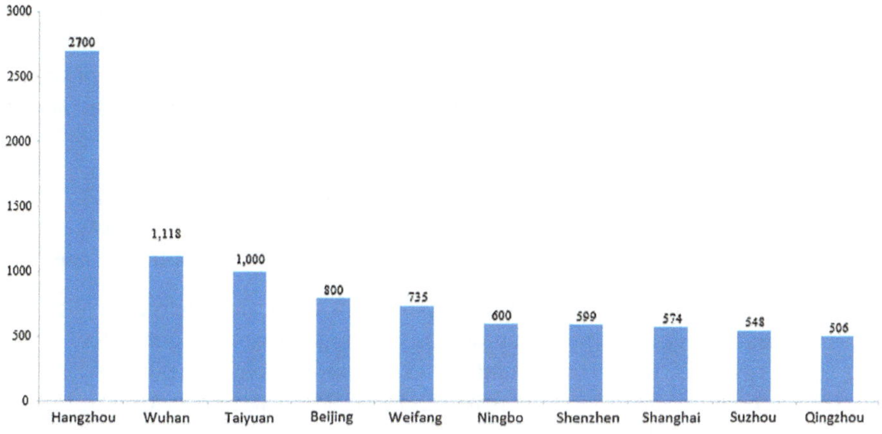

Fig. 8 Number of urban bicycle rental points for top 10 cities in China

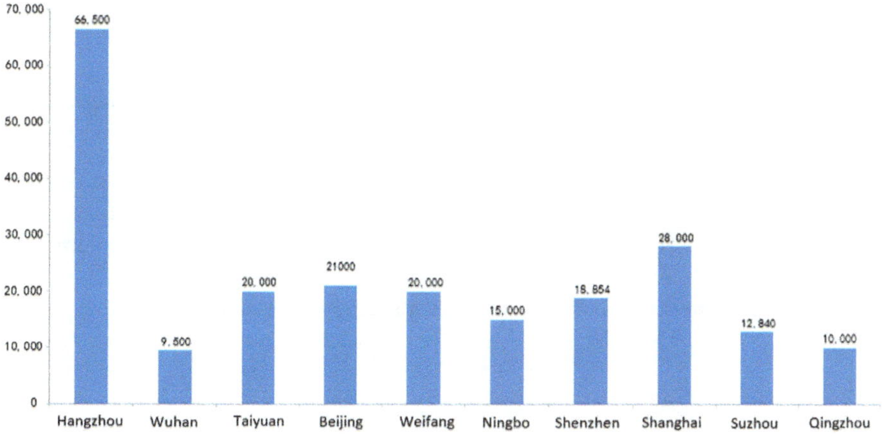

Fig. 9 Number of public bicycles for top 10 cities in China

4 Development Strategy and Policy Prospect of Transportation in China

The development strategy and policy prospect of transportation in China are mainly reflected in the following six aspects:

I. Insisting on development patterns of New-type Urbanization, and playing the guiding role of transportation system in urban development.

(a) Taking the construction of reasonable city patterns and promotion of the coordinated development of large, medium and small cities as the starting point, adjusting the development strategies of transportation, to better play the supporting and guiding role for urban development.
(b) Constructing compact cities, applying the transit-oriented development (TOD) strategies, curbing the disordered expansion trend of cities, and better leading the cities to develop compactly and orderly along the corridors of high-capacity public transportation.
(c) Taking spatial structure and the adjustment of industry layout as means, to optimize the function of cities and urban agglomerations, and to balance the public service resources.
(d) Establishing and completing the traffic impact assessment system.

II. Adjusting the transportation supply strategies, encouraging the low-carbon transportation, and giving priority to the development of public transport.

(a) Formulating classified guidance policy on traffic development patterns.
(b) Establishing the transportation construction investment guarantee system adapted for urbanization and social economic development process.
(c) Actively promoting the construction of rail transit, and making full use of its critical role in comprehensive transportation system of large cities and its guiding role for the development of urban agglomeration.
(d) Further promoting the development of conventional public transportation, and improving its accessibility and service level.
(e) Implementing strategies of priority for walking and bicycle, and improving traffic environment for pedestrians and cyclists.

III. Strengthening traffic demand management, and inhibiting the excessive use of private vehicles.

(a) Formulating differentiated policies of car ownership and use for different regions and different types of cities; giving play to the regulatory role of the market for allocation of resources, establishing and completing the traffic demand management system.
(b) Implementation of differentiated parking supply strategy.

IV. Paying attention to the planning and construction of integrated transport hub, and promoting the integration of the hubs and the city.

(a) Paying attention to the seamless connection of various transportation modes, and improving the level of the planning and construction of the integrated transportation hub.
(b) Optimizing the layout of integrated transport hubs, strengthening the integration of business, office and other functions in the hub and the surrounding areas, leading the adjustment of urban spatial structure.

V. Implementation of more strict policies for protecting environment and saving energy, and promotion of the application of advanced technologies.

(a) Promotion of clean energy vehicles and traffic modes with low energy consumption and low pollution, and reasonably raising vehicle emission and fuel standards.
(b) Promoting and improving the construction of urban transportation system of the smart city, improving the scientific level of traffic management and decision-making, and ameliorating the operation of transportation system.

VI. Strengthening policy guidance and perfecting coordination mechanism.

(a) Formulating the policy of differentially classified guidance according to the traffic development patterns.
(b) Establishing a strong mechanism for collaborative development of transport system in urban agglomerations.

5 Conclusion

Urbanization rate is still high and transportation will play a key role in economic development and will improve the life quality in china. Chinese society is continuously facing the serious transportation problems and challenges. The sustainable transportation system construction is our choice. We will improve the transportation service level by adjusting city form and land-use, giving priority to public transport, encouraging the use of bicycle and walk, improving the comprehensive transport hub, and changing the people's behavior. The key word is integration, which means integrating land use and transport, different transport modes, transport hubs and all planning together.

References

1. 2014 Urban Rail Transit Analysis Report (In Chinese).
2. China Public Bicycle (In Chinese). Websource: http://www.publicbike.net/
3. Statistical Bulletin of Transportation Industry Development of 2014 (In Chinese).

Civil Engineering Innovation Shaping Mobility of the Future

Boris Reyher and Mike Schlaich

Abstract Transportation infrastructure in the twenty-first century is faced with complex requirements. On the one hand, growing traffic volume worldwide puts increasing stress on existing infrastructure and requires upgrade and expansion of traffic networks. On the other hand, complex and diverging requirements such as economic efficiency, durability, sustainability, ease and speed of construction and last but not least, aesthetics are posed on the design and function of bridge structures today. schlaich bergermann und partner have been active in the field of bridge design for the past 30 years. Several noteworthy reference projects are used to discuss the role of civil engineers in solving the challenges posed on infrastructure design elegantly. These examples feature innovative concepts such as 'integral' design for railroad and highway bridges.

Keywords High-speed railroad bridge · Road bridge · Integral design · Engineering innovation · Bridge aesthetics

1 Introduction

Civil engineers' work today is faced with diverse and partly contradictory requirements which challenge their technical skills and experience as well as their creativity and ingenuity. In order to arrive at convincing solutions, engineers should always pursue a holistic approach. The aesthetic qualities of infrastructure are just as important when we recognize that like any other building work, civil structures are always also part of our *Baukultur*, a cultural statement.

After German reunification, a railroad bridge design council, the *Brueckenbeirat*, was established, which consists of engineers, architects and railway professionals, and

B. Reyher (✉) · M. Schlaich
Schlaich bergermann und partner, New York, USA
e-mail: b.reyher@sbp.de

M. Schlaich
e-mail: m.schlaich@sbp.de

M.C. Tandon and P. Ghosh (eds.), *Mobility Engineering*,
DOI 10.1007/978-981-10-3099-4_5

which discussed the aspects and principles of twenty-first century state-of-the-art railway bridge design. One result of the work of the *Brueckenbeirat* was a bridge design manual, the *Leitfaden Gestalten von Eisenbahnbruecken* [1], which aids engineers, architects and railway professionals in developing modern concepts, solutions and details for railway bridges. A selection of railroad bridge case studies designed and constructed following the principles and recommendations laid out in the *Leitfaden* are presented in the following, as well as an example for an integral highway bridge.

2 Case Study: Gaensebachtal Railway Viaduct

Gaensebachtal is a shallow valley located between cities Halle and Erfurt and required a low two-track viaduct of 1001 m length. The reference design following the standard German framework design foresaw a series of 4 m deep post-tensioned single span concrete box girders supported on structural bearings at concrete piers spaced at 40 m centres. Especially due to the low height of the viaduct above the valley floor, the reference design appeared visually dissatisfactory and inadequate.

To overcome the aesthetic problems, improve the durability and minimize the maintenance requirements for the structure, schlaich bergermann und partner conceived an alternative design, which was consequently put into practice [2]. A maximum of lightness and transparency was achieved first of all by reducing the maximum span to 25 m, which allowed for a reduction of the depth of the deck from 4 to 2 m. The viaduct is divided into 10 blocks, each a 112 m long integral post-tensioned concrete structure with a twin-web T-beam cross-section supported on twin concrete columns. The longitudinal braking and acceleration forces are resisted by characteristic V-shaped bracing walls located at the centre of each block. Similar bracing walls in transverse direction are located at the ends of each block to deal with transverse centrifugal forces (Figs. 1 and 2).

Fig. 1 Deutsche Bahn's reference design for Gaensebachtal viaduct; Schlaich Bergermann und partner's design concept for the same

Fig. 2 Completed Gaensebachtal viaduct; cross-section drawing; elevation drawing of two bridge blocks (*Photo* Adam Hoernig Baugesellschaft)

3 Case Study: Grubental Railway Bridge

Grubental is a short but rather deep valley located in the Thuringian mountains which required a 215 m long bridge to cross. Deutsche Bahn's reference design consisted of a concrete arch bridge with 60 m in the main span, flanked by two independent girder bridges on concrete piers. The bridge deck was to be supported on structural bearings throughout. schlaich bergermann und partner's alternative design [3] foresaw a concrete arch which fuses with the deck girder at midspan, creating a seamless and elegant structure. Next to the arch, slender integral concrete columns support the deck. Bearings are only used at the abutments, which is why the bridge can be referred to as a 'semi-integral' structure (Fig. 3).

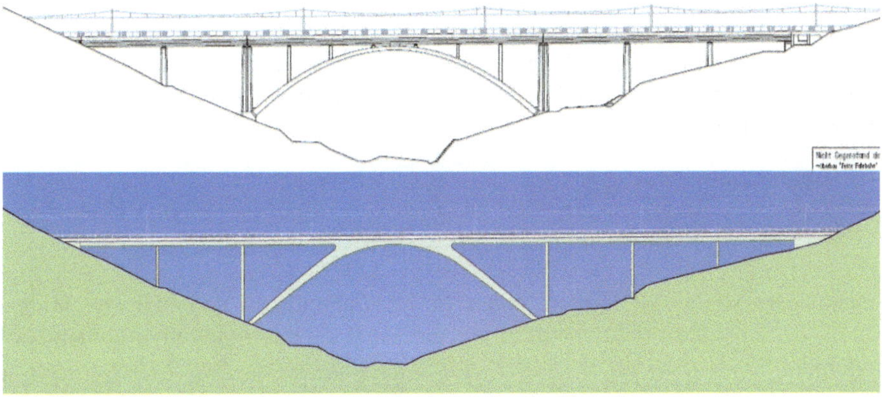

Fig. 3 Deutsche Bahn's reference design for Grubental bridge; Schlaich Bergermann und partner's alternative design concept

Fig. 4 Completed Grubental Bridge; cross-section and elevation drawings

The strut-like arch with its distinctive crown allows almost identical continuous spans of the superstructure, a twin-web T-beam with longitudinal and transverse post-tensioning. The tapered slender pier walls are designed to absorb the deformations due to temperature changes and shrinkage without bearings (Fig. 4).

4 Case Study: A11 Highway Viaduct, Belgium

The final case study deals with a four-lane highway viaduct in Belgium. The A11 motorway near Zeebrugge crosses a railroad and a shipping canal, requiring it to rise to 16 m above the flat landscape. The reference design by the Owner foresaw a series of conventional continuous post-tensioned concrete girders supported on bearings for the dual carriageway structure. schlaich bergermann und partner's alternative design for the 770 m long major bridge foresees a slender concrete deck supported on integral pier walls at 34 m centres. Only at the extreme two pier axis at either end, the structure relies on bearings to deal with the large temperature movements (Fig. 5).

5 Conclusions

Several case studies of integral and semi-integral railroad and highway bridges clearly demonstrate the advantages and potential of this innovative construction principle.

- The slender structural elements increase the elegance and transparency of the structure.

Fig. 5 Construction of A11 highway bridge (*Photo* Jan de Nul Group)

- Bearings and joints are largely omitted resulting in a reliable, robust and low-maintenance structure.
- Shorter spans allow solid cross-sections for the superstructure such as T-beams. Continuous girders without movement joints increase the efficiency of the superstructure in terms of its rigidity and damping behaviour.
- Slender integral piers with solid cross-section are more efficient to construct than the standard hollow box piers.

The integral method will prevail in many areas of railroad and highway bridge construction due to the favourable behavior of the structures, their efficiency, their economy, and their sustainable, holistic and aesthetic qualities.

References

1. J. Schlaich, V. Schmitt, S. Marx et.al.: "Leitfaden Gestalten von Eisenbahnbrücken", DB Netze AG, Berlin, (2008).
2. M. Schenkel, A. Goldack, J. Schlaich, S. Kraft: "Die Gänsebachtalbrücke, eine integrale Talbrücke der DB AG auf der Neubaustrecke Erfurt-Leipzig/Halle", Beton- und Stahlbetonbau 105 (2010), Vol. 9, Ernst & Sohn Verlag, pp. 590–598.
3. A. Keil, P. Wenger, T. Fackler, J. Schilling: "Die semiintegrale Grubentalbrücke – Besonderheiten bei der Planung", Bautechnik 88 (2011), Vol. 10, Ernst & Sohn Verlag, pp. 723–730.
4. M. Schlaich, "Integral Railway Bridges - New Bridges in Germany", ICI-IWCConference, Hyderabad, 2013.
5. M. Schenkel: "Verkehrsprojekt Deutsche Einheit (VDE 8.2): Eisenbahn- Neubaustrecke Erfurt-Leipzig/Halle – Herausragende aktuelle Baumaßnahmen", VDI-Bautechnik Jahresausgabe 2010/2011, pp. 41–51.

Future Shapes of Bridges in Urban Environment

Max Meyer

Abstract Improvement of physical mobility in cities is a big challenge. This paper investigates how approach to design and construction of elevated traffic corridors needs to evolve to meet the challenges of the future.

Keywords Aesthetics · Constructability · Carbon footprint

1 Introduction

Even though virtual mobility is becoming more and more important, enabling physical mobility in urban environment is and will remain a major necessity for cities to remain or become attractive places for living and for working. Due to lack of space new pedestrian, vehicular and rail links can often not be built at grade any more, but need to be moved below or above ground. Subject of this paper is to talk about the challenges with regard to building such links above grade on bridges focussing on the following three aspects:

- impact on a city's architecture
- impact on city dwellers during construction
- carbon footprint during construction and operation.

2 Aesthetics of Bridges

2.1. General: Bridges are huge structures, which in urban area are normally above at grade road corridors. Its visual impact on commuters at grade is huge. Bridges obstruct free view on surrounding buildings and the sky, restricting

M. Meyer (✉)
VSL International, Bern, Switzerland
e-mail: Max.Meyer@vsl.com

M.C. Tandon and P. Ghosh (eds.), *Mobility Engineering*,
DOI 10.1007/978-981-10-3099-4_6

sun exposure and growth of vegetation. All these comments show, that integration of bridges into an urban area in an architecturally satisfying matter is a demanding task. Since bridges are built to last for 100 years or more, they do impact urban architecture for decades.

2.2. Road bridges: Roads are either elevated only locally on flyovers to remove level crossings with other roads or elevated over extended length on viaducts to increase capacity of existing at grade roads. Flyovers are easier to integrate into surroundings than viaducts, since they are of relatively short length and have constant width. Intersections they cross are by definition wide open areas. Viaducts on the other hand have on and off ramps, where typical road section has to be significantly increased. Available road corridors are often narrow resulting in a scenario, where the elevated viaduct creates for the existing at grade road and the lower 2–3 storeys of buildings along the road a tunnel effect. Hence it may not always be advisable to increase vehicular capacity in existing corridors by construction of viaducts.

2.3. Rail bridges: When rails are elevated in urban areas they normally run over many kilometers on viaducts. These viaducts are typically only 9–10 m wide. Hence they are easier to integrate into existing road corridors when compared with road viaducts. The same can however not be said about the elevated stations, which are required at intervals of about 1 km along a viaduct. Stations are huge structures, which are significantly wider than the standard viaduct.

2.4. Pedestrian bridges: They are clearly the easiest to integrate satisfactorily into an urban environment.

2.5. Design and detailing: Once a decision has been made, that a bridge or a viaduct shall be built to increase mobility, it is the design engineer's duty to ensure, that the structure she/he is designing does blend well into the surrounding area and enhances its environment. This can be achieved with relatively small amount of extra costs, if basic aesthetic principles are observed already during the conceptual design phase.

   2.5.1 Deck: During conceptual design phase not only structural function of deck has to be looked at but designer needs to fully appreciate all the furniture the deck has to carry like parapets, noise barriers, drainage pipes, light posts (for road bridges) and catenary masts (for rail bridges). Only when looking at the entity of all these aspects an informed choice of the most suitable deck section can be made, which is structurally efficient, easy to build, economical and also aesthetically convincing. Another important consideration to arrive at aesthetically pleasing structures is to smoothen out any kind of transitions like e.g. variation in depth or width of the deck.

   2.5.2 Piers and portals: They are specifically sensitive with regard to aesthetics. Cross sections of piers and portals are therefore often controlled by aesthetic considerations and not by structural demands. Pier need frequently to be widened towards the top in transverse direction to achieve a smooth interface with the deck.

2.6 Concrete: As explained below concrete is worldwide the predominant material for construction of bridges. The beauty of concrete is that it can be built to almost any shape due to its moulding characteristics. Hence it is a very good material to smoothen out transitions as mentioned above and it allows incorporation of architectural features into external surfaces of for example piers and parapets.

3 Impact During Construction

3.1 General: Construction of a bridge or viaduct is not a matter of days but a matter of 2–3 years. Construction happens along existing road corridors, which are generally jammed by cars and people. Hence it is imperative to select structural concepts, which allow to move as much of the construction activity off site. Concrete is worldwide the predominant material for construction of bridges due to fact that main components are almost always locally available and maintenance is minimum if design, detailing and construction are correctly carried out. Technologies are from precast segments, which span over the full width of the deck and are held together with longitudinal prestressing tendons. This eliminates need for any false work and removes almost completely need for casting of concrete on site.

3.2 Deck construction: Even so the concept of moving as much work off site by using precast segmental technology is understood by owners, consultants and contractors, the potential this technology has with regard to shortening construction time and reducing negative impact during construction on people commuting and living along the bridges being constructed, is often not exploited. Reasons for this situation are:

3.2.1 The design and detailing of the deck is done without consideration of construction aspects. In order to be able to do that, designer needs to have in depth understanding on how segments of precast segmental viaducts are being produced and erected.

3.2.2 Contractual arrangements are not adapted for this technology: Precasting is all about standardization and mass production under factory conditions. As mentioned above the designer needs to have in depth understanding on how segments are being produced and erected. Therefore often owner goes for design/build type contracts. This may be fine if design and construction of an entire network is given to one contractor, but has more disadvantages than advantages when design and construction is broken down into several contracts, as it is normally the case. The owner of a network ends then up with different structural details with regard to bearings, joints, drainage systems etc. making the maintenance a nightmare both with regard to management and costs. So engagement of an experienced and reputed consultant directly by the owner to carry out design/detailing for a standardized deck, which is used for the entire network, appears to be a more sensible approach.

3.3 Piers: Piers like decks be made up of precast segments, which are joined together by vertical prestressing tendons.
3.4 Foundation: Foundation schemes, which have limited dimensions in transverse direction, are preferred since viaducts are often going along central dividers of existing at grade roads.
3.5 The future: The industry is far from utilizing the full potential of bridge decks made up completely from match cast precast segments. A new mind set at owner and consultant side is required to improve the situation.

4 Carbon Footprint

4.1 General: we need to differentiate between carbon footprint during construction and carbon foot print during operational phase of a bridge.
4.2 Construction:

4.2.1 Carbon foot print can be reduced in construction by selecting materials, which have a small carbon foot print during manufacturing, reducing need for transport of material and cutting down on energy consumption during construction.
4.2.2 To reduce carbon footprint for concrete bridges is to prestress the concrete, since this allows to reduce average concrete consumption per m2 deck and to reduce steel consumption.
4.2.3 Precasting deck and may be even piers requires standardization to allow mass production of big part of the bridge under factory conditions. This will help to lower carbon footprint, since mass production under factory conditions is more energy efficient than craftsman type production on site.

4.3 Operational phase:

4.3.1 Reduction in carbon footprint during operational phase is achieved by reducing need for maintenance of the bridge. Main structural elements of bridges in prestressed concrete, if correctly designed, detailed and built, require minimum maintenance. However that is not true with regard to bearings and joints. Therefore bridges should be designed as much as technically feasible and sensible with deck continuity over several spans and with frame action, since such structural systems are characterized by minimum number of bearings and expansions joints.

4.4 The future: Designs need to become sharper to minimize quantities of permanent work material and need to go as much as technically feasible and sensible for deck continuity and frame action.

5 Conclusion

The title of this paper may have created expectations that the future of bridges will all be about new bridge concepts. Of course nobody is able to predict the future. The author's opinion is however, that "conventional" bridges still have a place in the future to improve physical mobility in cities, even so upgrading of capacity of transport corridors will have to go more underground. The predominant material for the decks of the bridges will probably remain prestressed concrete due to its high local content, its moulding characteristics and its relatively overall low carbon footprint during construction and operation phase. However major changes to mindsets of owners and designers building bridges in concrete in urban environments have to take place to fully exploit the potential inherent in prestressed concrete and in particular in precast segmental prestressed concrete. Owners have to choose suitable contractual models and designers need to have a holistic approach going well beyond structural integrity checks only but considering as much aesthetic aspects, constructability, material consumption and future maintenance demands [1].

Reference

1. Max Meyer, "State of the art bridge construction - the challenge to choose the right method", IABSE Symposium BKK 2009.

Making of Bandra-Worli Sea Link

Santosh Rai

Keywords Precision engineering · BWSL · Urban mobility · Sea bridge construction · Superstructures

A TRIUMPH OF PRECISION ENGINEERING

S. Rai (✉)
HCC, Mumbai, India
e-mail: Santosh.rai@hccindia.com

M.C. Tandon and P. Ghosh (eds.), *Mobility Engineering*,
DOI 10.1007/978-981-10-3099-4_7

1 Introduction

The Bandra-Worli Sea Link (BWSL) is a civil engineering marvel spanning an arc of the Mumbai coastline. With its cable-stayed towers soaring gracefully skywards, the sea link is a reflection of the modern infrastructure that Mumbai is adding in its progress towards becoming a world-class city.

The BWSL project is a part of the Western Freeway Sea Project, which, in turn, is a part of a larger proposal to upgrade the road transportation network of greater Mumbai. In the first phase it will connect Bandra to Worli whereas in the subsequent phases the plans are to take it further to Haji Ali and then to Nariman Point.

The project has been implemented by the Maharashtra State Road Development Corporation Ltd (MSRDC) and the Maharashtra Government and is being built by HCC (Hindustan Construction Company). The BWSL project offered HCC an opportunity to accomplish one more feat: to construct an eight-lane freeway over the open sea for the first time in India.

2 Highlights in Brief

- India's first bridge to be constructed in open-sea conditions.
- 4.7 km, twin, 4-lane independent carriageway bridge across the open sea.
- 16-lane toll plaza with 20-m wide promenade together with state-of-the-art traffic monitoring, surveillance, information and control systems.
- 2342 pre-cast segments for total bridge with varied width.
- 40,000 MT of reinforcement, 23,0000 cum of concrete, 5400 MT of Post tensioning strands and bars used.
- Osterberg cell technology used for the first time in India to check pile strength (for up to 9600 MT).
- Engagement of Asian Hercules, one of the largest floating shear leg crane in the world for shifting 1260 MT launching truss from Bandra end to Worli end of the main cable stay bridge.
- Largest span for cable-stayed bridge in India.
- Up to 25-m high pier in open sea, giving ample headroom to marine traffic.
- Use of Polytron Disc in bearings on piers for the first time in India.

3 Challenges Encountered During Execution of the Project

Following were the major challenges to be addressed:

- The foundations of the bridge included 604 large diameter shafts drilled to lengths of 6–34 m in geotechnical conditions that varied from highly weathered volcanic material to massive high strength rocks.

- The superstructure of the approach bridges were the heaviest spans in the country to be built with span-by-span method using overhead gantry through a series of vertical and horizontal curves.
- A one-of-its-kind, diamond shaped 128 m high concrete tower with flaring lower legs, converging upper legs, unified tower head housing the stays and a throughout varying cross section along the height of tower.
- Erection of 20000 MT Bandra cable-stayed deck supported on stay cables within a very close tolerance of deviations in plan and elevation.

3.1 Foundation and Substructure

The foundations for the BWSL project consist of 2000-mm diameter piles numbering 120 for the cable-stayed bridges and 1500-mm diameter piles numbering 484 for the approach bridges. The working load on the approach piles ranges from 700 to 1500 tons whereas for the piles below the cable-stayed bridge working load is 2500 tons. The major engineering problems that needed suitable solutions before proceeding with the work were as follows:

1. Highly variable geotechnical conditions of the foundation bed as explained above.
2. Highly uneven foundation bed even for plan area of one pile.
3. Presence of Intertidal Zone (Foundation Bed exposed in low tide and submerged in high tide).

For conducting the load test on the piles, the load to be applied varied from 4500 to 9600 tons. Arranging reactions for such loads either by normal kentledge method or by soil anchor required massive scale arrangements in the sea waters. This was completely avoided by a careful planning of load test using the Osterberg load cell method (Refer Fig. 1).

Reverse Circulation Drilling method is adopted for foundation construction. The highly uneven foundation beds and the presence of intertidal zone brought in lots of difficulty in terms of Liner pitching. This problem was solved by constructing a gabion boundary at the bed level around the casing, pouring concrete between the casings to make an artificial penetration of the casing. After setting of the concrete under the water, drilling was commenced using RCD.

1. **Superstructure**

The BWSL Project has (9 + 2) approach bridge modules. These modules range from 3 continuous span units to 8 continuous span units. The deck of the carriageways consists of triple cell precast box girders supported on piers founded on independent substructure. The Concrete Grade for the superstructure is M60. The average weight of the span is 1800 tons, whereas the heaviest span in the bridge (to

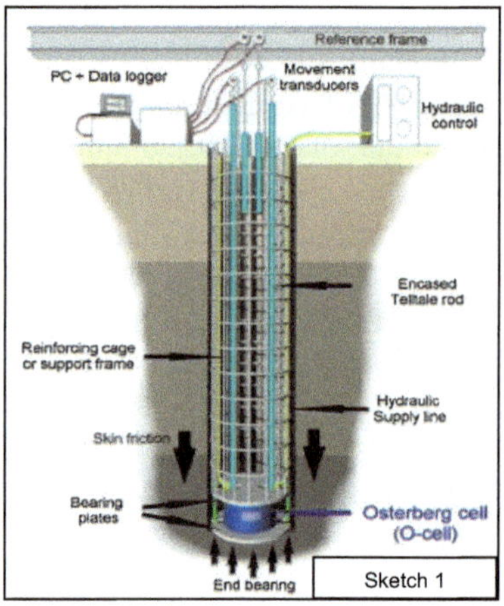

Fig. 1 Load testing of piles by osterberg load cell

be erected with the Launching Gantry) weighs 2000 tons. In addition, the trusses were to be designed to receive the segment from the already erected deck as well as from barges parked directly under the truss. The erection gantry is 1260 MT truss designed to erect spans for the above configuration. The unique feature of the truss indeed is the maximum span weight it can handle and that it can launch the pier and EJ segment itself. The truss also has the capacity to align the total span in hanging condition after the gluing is completed. The truss is fully mechanized for self launching and aligning. An individual segment can be aligned on the truss using a set of four hydraulic jacks mounted on each suspension frame. In order to eliminate the casting or erection errors within a span, two wet joints are provided on either end of the span. The wet joints are cast after finalization of the span alignment.

3.2 Relocation of Launching Trusses Using 1600 MT Capacity Barge Mounted Crane—Asian Hercules

After erection of the deck on Bandra side, the trusses were required to be shifted across the Bandra cable stay bridge by 600 m to Worli side to take up the spans beyond the Bandra Cable Stay. Various options like (i) dismantling of the trusses at present locations and reassembling them at new locations, (ii) lowering the trusses on a suitable floating craft and shifting and erecting them, and (iii) shifting the total truss using a floating crane, etc. were analyzed in detail. Taking into consideration various constraints the best option available was the relocation of the trusses in one piece using a floating crane.

Asian Hercules is one of the biggest floating shear leg cranes in the world. This crane is mounted on a barge which is over 240 feet long and more than 130 feet wide, weighs 5900 tons and has enough lifting capacity (1600 MT) to lift a weight equal to 2000 small cars.

After lifting the truss, Asian Hercules crane was moved to a safe location where enough water depth (draft) was available to park the crane with the truss. Then the crane with the truss was moved to a new location during the next high tide. The truss was then lowered on to the final location. The lowering of truss at the final location was achieved through Guides, which helped to achieve a final placement accuracy of ±50 mm. The operation, which otherwise would have taken one complete year, was completed in matters of a few days.

4 Cable Stay Bridges

It is for the first time that cable stay bridges have been attempted on open seas in India. Coupled with the fact that the aesthetically designed pylons have an extremely complex geometry and one of the longest spans for concrete deck, the challenges encountered were indeed formidable (Refer Figs. 2 and 3).

5 Construction of Pylon Tower Legs

The salient characteristics of the pylon tower that make it complex and challenging from the point of view of constructability are as follows:

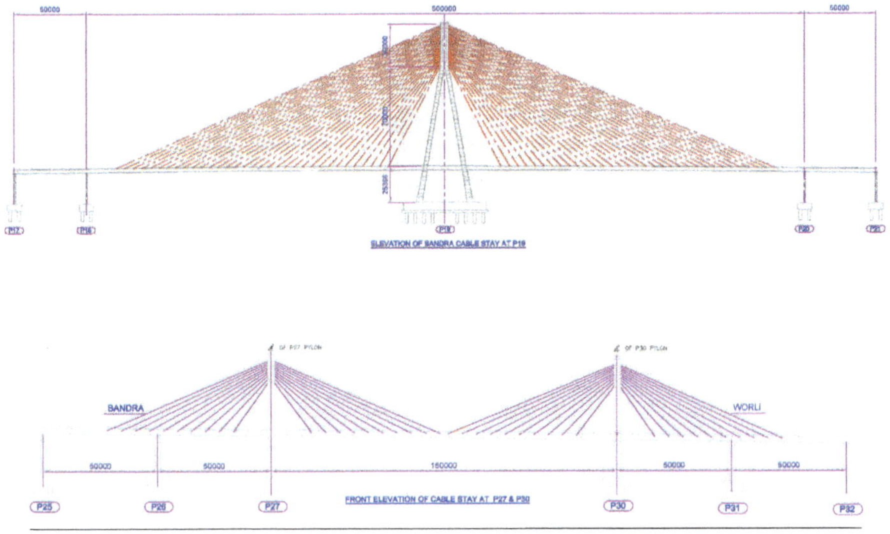

Fig. 2 **a** Bandra cable stay bridge **b** World cable stay bridge

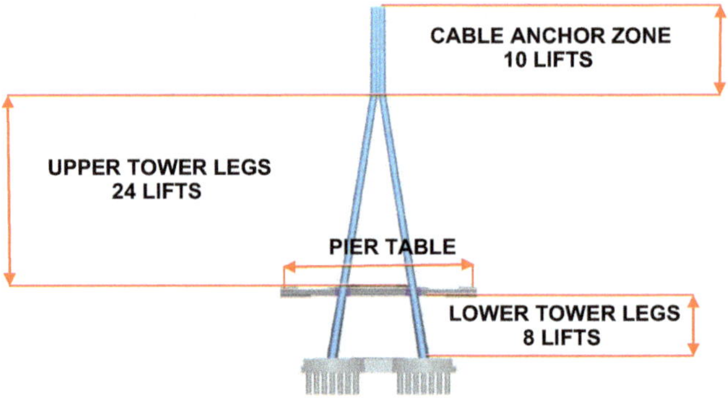

Fig. 3 Side view of tower P19 for bandra cable-stayed

(a) The tower legs are inclined in two directions and section decreases gradually with height;
(b) Construction joints permitted only at 3 m level. Inserts were permitted only in horizontal grooves provided at 3 m height.

To start with, the project design team designed an automatic climbing shutter formwork system, which was fabricated on site and employed to execute all tower leg lifts below deck level. To affect further reduction in time cycles, HCC approached Doka, Austria. Doka then devised a customized solution based on their SKE-100 automatic climbing shutter system (Refer Figs. 4 and 5).

(a) **Survey of Tower Legs**

The complex pylon geometry was another challenge for surveyors. Coupled with geometry, the construction stage analysis indicated leaning and progressively increasing inward inclination of pylon legs during construction. HCC's Principal Surveyor devised a sophisticated technology to measure coordinates through a combination of total station and prisms mounted on pylon legs. The temperature and construction stage analysis factors were applied to derive the corrected

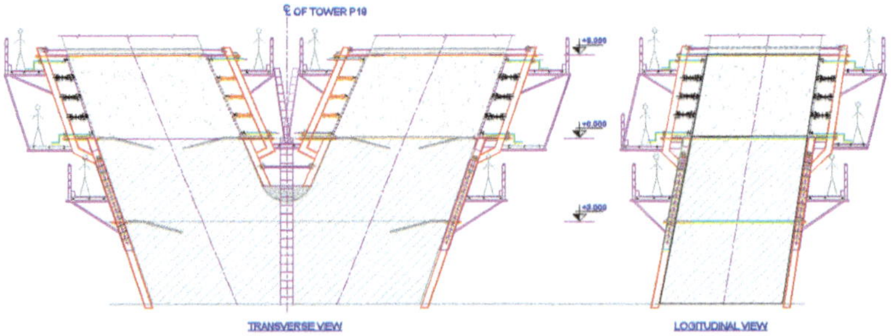

Fig. 4 Construction of tower legs below deck level

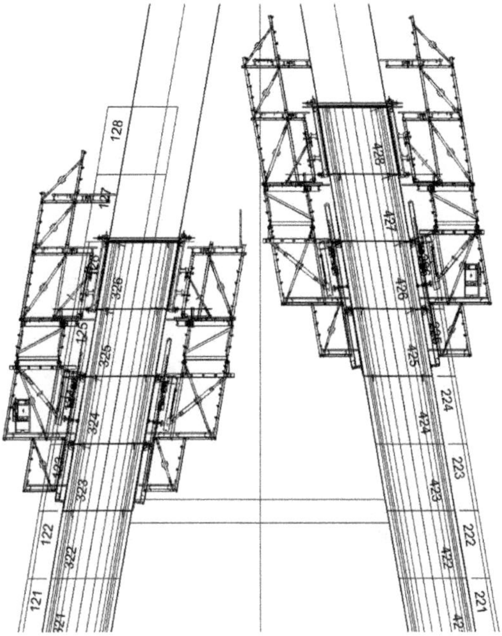

Fig. 5 DOKA SKE-100 Automatic Climbing Scaffolding System Erected on Tower Legs

coordinates. The pylon legs were constructed within an accuracy of ±5 mm, which speaks volumes about the technique employed.

(b) **Anchorage Box**

Anchorage Box is used as inner shuttering for tower head. Bearing plates with guide pipes are fixed to the anchorage box. Guide Pipe and Bearing Plates actually transfer the deck loads to tower concrete which are generated due to stressing of stay cables. The anchorage box is fabricated with 12 mm thick high grade steel plates. It is fabricated in pieces and then bolted at tower head portion. The bearing

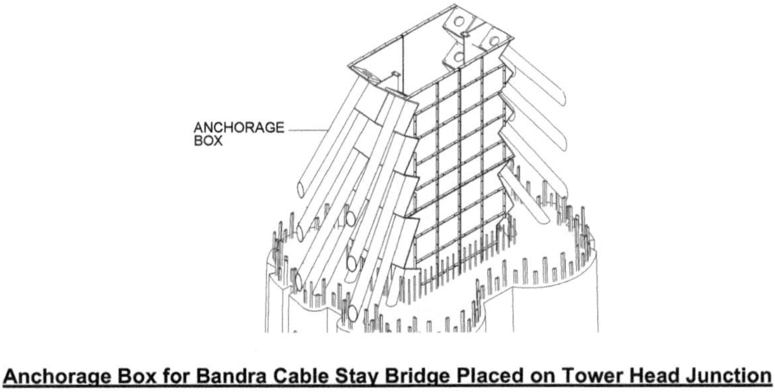

Fig. 6 Anchorage Box for Bandra Cable Stay Bridge Placed on Tower Head Junction

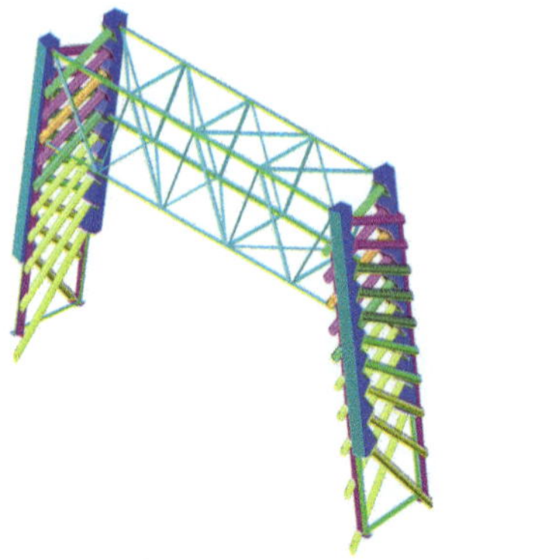

Fig. 7 Anchorage Box for Worli Cable Stay Bridge

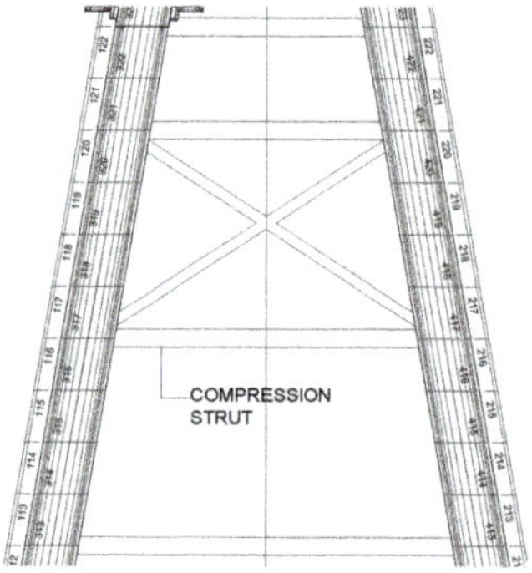

Fig. 8 Temporary arrangment for controlling geometry during construction

plates and guide pipes of anchorage box are galvanized and the remaining portion was painted with anticorrosive polyurethane based paint. Anchorage boxes are fixed with the help of co-ordinate system for accurately fixing the anchorage point and angle of stay cable.

(c) **Compression Struts**

Compression struts are provided to maintain the precamber and stresses during construction in their required order.

(d) **Erection of Segments of Cable Stay Bridge by Derrick in Sea Waters**

The method used for erection of segments at Cable-Stayed Bridge was balance cantilever construction method. During construction, the length of free cantilever for Bandra Cable-Stayed Bridge was 215 m and for Worli Cable-Stayed Bridge it was 73 m.

The segments were lifted by the instrument named Derrick which was fixed on both ends of the pier table segment and then forwarded. Lifting operation was done simultaneously on both ends. At a time, Derrick can lift one segment. Deck is

Fig. 9 Lifting of Segment with Derrick

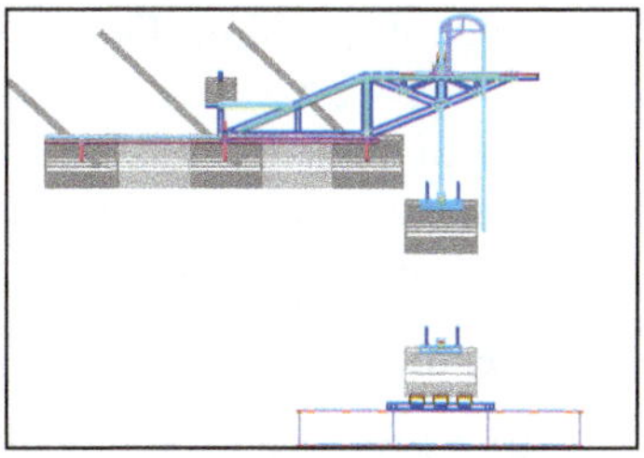

constructed of alternate stay and non stay segments joined to pier table segments. The challenge was in picking the segment from the barge amidst tidal conditions and under current. Any major movement of barge with the segment load still on barge could bring serious damages to the cantilever deck. A detailed scheme for segment lifting and release from the barge was developed.

1. **Working During Monsoon**

The Maritime Board does not allow marine traffic in monsoon season. Thus, work was halted mid-May only to re-commence in October, effectively reducing the work schedule to only seven months in a year. To overcome this hurdle HCC put forth the solution in the form of an innovatively designed temporary bridge. This bridge had a total length of 325 m. It had the facility of a walkway, a concrete pipe line, an electrically-operated trolley mounted on rail, water line and a pipe line. It paved the way for successful continuation of work during the monsoon season when the sea was rough and the winds were strong.

2. **Logistics**

Another challenge was ensuring effective supply chain at all working locations spread across the alignment in the sea and formulating measures to ensure the same. State-of-the-art electronic devices were placed on the barges to cut down on idle timings. Navigation and transporting 19 precast segments in 24 h at different open sea locations was a challenge. Secondly, concrete consumption at the peak had been at the rate of 50 cum/h. Under marine conditions, the consumption rate has been in the order of 700 cum per day. To add to this, maintaining adequate food supply for around 2500 people (in a shift) working in the sea at over 30 locations was a big challenge. These complete requirements were met with an effective utilization of a fleet of 30 marine vessels including 13 barges for concrete, segments and material transport, eight steel boats for material and workers transport, three tug boats and six smaller passenger boats.

Around four passenger boats were used for carrying food to approximately 30 locations in the sea. Each employee, while starting his day, entered the log indicating the location at which they would be working. Thereafter began the clockwork of gathering tiffin boxes, washing and cleaning, allocation and dispatch as per the log entries along with the drinking water supply including tea supply at two time intervals per shift. During rough sea conditions in the normal working season, extra tiffins were carried to take care of possible spillage while transferring the tiffins from boats to working locations. Thus workers were also suitably cared for, while meeting the engineering challenges posed during construction of the Bandra-Worli Sea Link.

3. **Psychological Conditioning**

Lack of awareness is the biggest hazard for safety. Since the primary safety hazard are related to engineering control, equipment, job methodology, material handling, structural fabrication and emergency preparedness, HCC made sure that every worker is taken through the HSE program. The orientation program made them aware of the various safety hazards associated with a project and necessary precautions to be taken to prevent them. They are also taught how to evacuate during any emergency. For its meticulous planning and implementation of safety practices for the BWSL project, HCC has won the prestigious "Golden Peacock Award" for safety, health and environment in June 2007.

6 Conclusion

The Bandra Worli Sea Link which incorporates the largest cable stayed bridge span in India is truly a remarkable structure. Constructed in open sea conditions using precast segmental construction technology for the superstructure and large diameter piles for the foundations it employed heavy floating cranes and other equipment in its realisation. Installation of piles in uncertain geotechnical conditions consisting of highly variable rock was one of the major challenges that had to be overcome. The handling of launching truss of 1260t at different locations was another.

Rough sea conditions reduced the construction operations in the sea to a mere 7 months in a year. The logistics of catering to 2500 workmen in one shift at same 30 locations along the alignment presented its own problems. The safety, health and environment (SHE) aspects maintained an excellent record throughout the construction period.

Digitally Reimagining Mobility

K. Ananth Krishnan

Abstract Being mobile is natural to man. Engineering technologies have contributed immensely to human mobility. This paper looks at how current computing technologies are impacting human mobility and sustainability. We present a few examples from the TCS research standpoint, how digital forces improve the travel and transportation ecosystems in an incremental manner. We also look at some disruptive changes digital forces can bring to the travel and transportation industry. We discuss some examples of digital forces that can contribute to sustainability.

Keywords Mobility · Digital · Incremental innovation · Disruptive innovation · TCS research

1 Introduction

Knowledge and communications technologies have played a role in each leap mankind has taken in terms of mobility. If the compass was engineering's contribution to travel, then maps were a "knowledge" contributor. The printing press was a fundamental contributor to disseminating scientific learning during the Renaissance and catalyzing the following Industrial Revolution, as it helped cumulative knowledge growth. If the large steam ships of the 19th Century made passenger and cargo transport faster and cheaper, the telegraph was making travel safer and providing more business value for traders.

The twentieth century saw the conquest of the skies, with commercial airlines and space travel. It also saw large-scale growth of computers, which enabled computing, information and communications technologies to converge. This resulted in safer and cheaper long-distance travel, enriching the human mind and spurring globalization. For example, during World War I, it was discovered that flying for transportation and reconnaissance was not effective if the crew could not get in touch with their col-

K.A. Krishnan (✉)
Tata Consultancy Services, Mumbai, India
e-mail: ananth.krishnan@tcs.com

M.C. Tandon and P. Ghosh (eds.), *Mobility Engineering*,
DOI 10.1007/978-981-10-3099-4_8

leagues on the ground. AT&T provided the first air-to-ground and ground-to-air radio communications.[1] Today's space programs make extensive use of computing and communications technologies.

Thus we see through history, both engineering and communication technologies being associated with advancements in modes of travel and transportation, as well as the dissemination and exchange of information across all the stakeholders of the Mobility ecosystem.

Against this backdrop, this paper looks at how current computing technologies are impacting human mobility and sustainability. We present a few examples from the TCS research standpoint, on how these digital forces improve the travel and transportation ecosystems in an incremental manner. We also look at some disruptive changes digital forces can bring to the travel and transportation industry. We discuss some examples of digital forces that can contribute to sustainability.

2 The Five Digital Forces

TCS believes that five computing-technology related forces are driving big changes in industries, within businesses and in society, at large. We have termed these the Digital Five Forces:

1. Mobile and pervasive computing
2. Social networks
3. Cloud computing
4. Big Data and analytics
5. AI & robotics

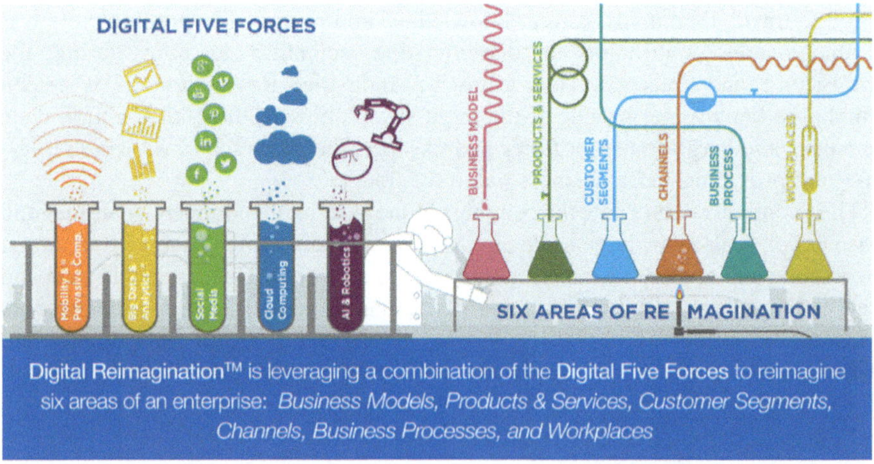

[1]http://www.corp.att.com/attlabs/reputation/timeline/17air.html.

TCS believes that each of these forces is impactful by themselves, but a combination of the forces can reimagine business models, products and services, customer segments, channels, business processes, and workplaces in any industry.

These forces are driving incremental as well as disruptive changes in the human mobility ecosystem.

3 Incremental Innovations in the Mobility Ecosystem

Many engineering, industrial services, travel and transportation and civil engineering sectors have been deploying computing power. The maturing of digital technologies have created a greater capacity to innovate in core areas, improve internal processes, products and services.

3.1 Manufacturing of Vehicles

Every aspect of the automotive industry, from design to new product introduction, involves a range of computing technologies. Manufacturing and shop-floor activities are going digital. CFD simulations (power-train thermal, air flow simulations), structural simulations (Noise–Vibration–Harshness (NVH), durability, impact simulations, and design sensitivity) as well as embedded systems development and testing, enabled by the Digital Five Forces, help bring better vehicles to market. Intelligent manufacturing (enabled by the Internet of Things) not only allows for seamless production, but also offers proactive maintenance, ensuring safety and efficiency of vehicles.

3.2 Infrastructure Engineering

The construction industry is growing in complexity with ambitious projects and introduction of new materials. Civil works such as bridge building are ever more reliant on 2D/3D and even 4D structural modeling using big data and cloud infrastructures allowing for factoring in wind dynamics, earthquakes, and wear and tear across a time period.

3.3 Traffic Management

Several aspects of traffic management from scheduling of trains to tracking and modeling road traffic to driver behavior monitoring are being improved by digital technologies.

Modelling and simulation (M&S) techniques play an important role in intelligent transportation systems, and enable traffic planners to predict traffic parameters of interest as well as to study and analyze the impact of planned and unplanned changes (lane closures, evacuation scenarios, inclement weather conditions, and so on). However, it is important to complement empirical models with data driven models. The GPS hardware present in smartphones, coupled with cheap and ubiquitous connectivity to the network, serves as a near-ideal data collection device, at a scale that is large enough to capture the full richness and diversity of the data enabling better analysis.

3.4 Logistics and Scheduling

In large transportation networks, such as a nation's rail system or airways, operational efficiency on the ground is critical. Scheduling, for example, is an area that requires optimal planning. The Indian railway network is one of the largest and busiest rail networks in the world, It has over 10,000 stations and over 10,000 concurrent trains. Train dispatches across different territories and tracks are based on human decision making. Introduction of new trains, cancellations, addition of temporary services and delays add huge complexities.

Given the advances in CPU, communications bandwidth, storage capacity and computing speeds, the numbers and sophistication of sensors, and in the sciences of AI and mathematical optimization, it is possible to automate several decisions. It is therefore feasible to monitor and reactively plan such large logistics and transportation networks to deliver significant economic and environmental benefits.

3.5 Driver Behavior Tracking

Human decision-making in personal transportation is also being revolutionized by digital interventions. Driver behavior is one such area. This is of interest to several stakeholders, like city councils, auto manufacturers, insurance businesses, and of course drivers. Several insurance companies are piloting smartphone apps for influencing vehicle drivers to follow safe driving habits to reduce the risk of accidental damage and consequent claims. This is applicable for both passenger as well as freight transport vehicles. The interesting fact is that the smartphones host different types of sensors like GPS, accelerometers, and gyroscopes in a single low-cost platform. Additionally, the computing power of modern phones has increased manifold. This offers us the opportunity to identify novel techniques of multi-sensor fusion so that the prediction accuracy is vastly improved.[2]

3.6 Goods Movement Through Supply Chain Efficiencies

There is a digital transformation happening across the entire supply chain. Apart from now commonly used bar code and RFID technologies, several newer technologies such as Bluetooth, ZigBee, iBeacons, NFC, WiFi, and smartphones are shaking up the supply chain landscape.

Traditional services that are now turning digital include metering, monitoring, and tracking, which offer enhanced in-transit visibility and pervasive visibility. Advanced services include diagnostics and prognostics (including detection of outliers, anomalies, faults, and failures), predictive maintenance, proactive replenishment, and connected supply chain solutions.

4 Disruptive Innovations in the Mobility Space

Clayton Christensen, Professor Harvard Business School, terms innovations that create completely new products, or products serving new target markets at a different price point "Disruptive".[3] We present some examples of how digital technologies are disrupting the human mobility space.

4.1 Preparing for Autonomous Vehicles

When autonomous vehicles are ready for commercial use, they will disrupt several processes and business models related to mobility. In the meantime though, new challenges surface. Deployment of such robots (that is, self-driving cars) on roads will gradually increase. Since these robots play by the book (and humans don't!), how do we, in the interim, overcome this dichotomy?

The possible neat solution to this peculiar problem is: 'blend' them. For example, crunching large amounts of driver behavior data and using AI we first identify an individual's naturalistic driving styles as that belonging to one broad category like cautious drivers. Then we can optimize the driving style until a human driver feels comfortable with the automated driving; keeping in mind, though, that

---

[2]Chowdhury et al. [1].

[3]http://www.claytonchristensen.com/key-concepts/.

safety and reduced fuel consumption is the primary goal of such an exercise. Finally, implement such anticipatory driving styles in the vehicle's Adaptive Cruise Control so that the human driver has enough trust to relinquish control the to machine. The issues related to control sharing between man and machine have become emerging areas of research challenges.

4.2 Disruptions in Usage Models

Even as we speak disruptions are happening in the way taxi services are being used, largely facilitated by mobile apps that mash up several other online services like maps and GPS systems: Uber and Lyft's ride sharing and renting systems are an example. If these models grow in developing countries without compromising passenger safety, they can lead to reduction of carbon footprint and congestion without compromising on comfort.

4.3 Lightweighting Automobiles

The need for enhanced fuel economy for reduced impact on the environment and also the need for enhanced safety is posing significant challenges to the automotive industry. For example, the automotive industry in the US has targets for increase in fuel economy to 54.5 miles per gallon for cars and light-duty trucks by 2025.[4] This degree of enhancement requires the industry to move toward electrical transmission as well as lightweighting. However, lack of availability of the right materials poses a challenge to achieving this goal. For electrical vehicles, materials used for energy storage in batteries are not adequate; light weighting requires better structural materials that reduce the weight of the vehicle, in spite of the increase in weight due to batteries.

Development and deployment of new materials has traditionally been an iterative process, with long lead times. Recent advances in computing infrastructure, mathematical modeling and simulation, and the enabling digital capabilities have a significant potential to bring about a paradigm shift in materials engineering. The discovery of new materials for needs such as energy storage has been accelerated by initiatives such as the 'Materials Genome Initiative', which envisages significant use of data mining, image processing, and machine learning techniques to accelerate material discovery.

The discipline of Integrated Computational Materials Engineering (ICME) enables simulation driven development and deployment of new lightweight materials, such as

[4]https://www.whitehouse.gov/the-press-office/2012/08/28/obama-administration-finalizes-historic-545-mpg-fuel-efficiency-standard.

advanced high strength steels, in a collaborative environment with close association of material scientists, manufacturing engineers, and product engineers.[5] Besides physics-driven modeling and simulation, the application of machine learning and knowledge engineering to materials engineering and high-performance computing is expected to play a significant role in this regard. Finally, such a collaborative development effort would require a significant IT-driven collaborative environment.

4.4 High Performance Computing—The Enabler

As we plan for better vehicles and better infrastructure, simulation and computational fluid dynamic modeling will become critical. To simulate scenarios such as traffic congestion, damage to transportation infrastructure, crash analysis for safety engineering of vehicles and visualization of mechanical problems, huge computing power is required. A typical CFD simulation exercise may take hours to complete, given hundreds of CPUs. High Performance Computing (HPC) technologies are the only means to satisfy these requirements. (HPC refers to a specific collection of hardware and software technologies that can together deliver large computational power required by applications.)

5 Digital Technologies, Mobility, and Sustainability

The World Business Council for Sustainable Development's Mobility Report 2030[6] that projects trends from 2000 to 2030 observes, "Personal and goods transport activity will grow rapidly, driven primarily by rapid growth in real per capita income." More money and more leisure apparently means more mobility.

This leads us to the question of why people travel. Forced movements (due to war, trafficking, and natural disasters), unfortunately, still contribute to population migration. Aspirational travel happens along the eternal themes of quest for better resources and living conditions, access to knowledge, and just the experience of travel. In each of these areas, digital technologies are playing an increasing role in reimagining mobility and increasing sustainability.

In today's globalized economy, businesses seek a global footprint. This increases the need for business related travel. However, if businesses invest in good communication infrastructure (messaging, voice, video, immersive experiences) business travel can be reduced. (For instance, TCS' sustainability policy mandates a

[5]Pollock and Allison [2].

[6]**Mobility 2030:** Meeting the challenges to sustainability published by The World Business Council for Sustainable Development (WBCSD). Full report, Page 28 Link: http://www.wbcsd.org/Pages/EDocument/EDocumentDetails.aspx?ID=69.

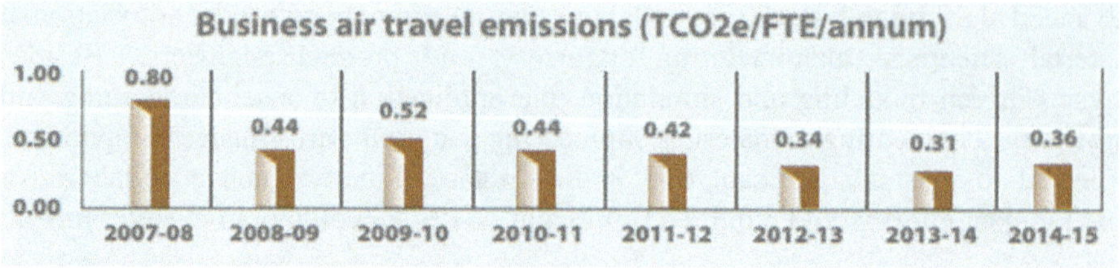

Fig. 1 TCS' carbon footprint relating to air travel

reduction in business travel. By improving our communication systems we have achieved a 55% reduction in per capita air travel over the base year 2007–2008. See Fig. 1, from TCS' Sustainability Report.[7])

Travel for day to day transactions such as shopping, banking, and bill payments is being substituted by digital processes. Some statistics that indicate this:

- Amazon declared that more than 2 billion units were ordered through its website in 2014; more than 25% of payments through Amazon Payments was on mobiles.[8] Flipkart, eBay, and other retail ecommerce sites worldwide reported hearty sales figures. We can safely assume several million shopping trips were reduced.
- An August 2015 report claimed that approximately 55% of Americans use mobile banking.[9] Many countries report similar online banking adoption rates. Kenya is a pioneer in branchless banking, with customers serviced only by mobile banking. So, fewer people in most countries are visiting banks for basic transactions.
- Electronic bill payment systems are overtaking over the counter systems in most parts of the world.

In its transport related equity concerns, the Mobility 2030 Report[10] raises the issue of the elderly, especially in countries like America, continuing to drive. It points out that their ability to drive safely will decline, and at this point their mobility will be curtailed. This poses a problem, since they are likely to need greater medical attention as elders. Remote monitoring through sensors allows non-intrusive health tracking and oversight. This allows caregivers to attend to the elderly as and when they need attention, without requiring 24-h intrusive monitoring. This again leads to a lot of saved trips both for the elderly and for the support community.

Work progresses on mobile based 'bio-sensing' to measure a number of physiological and behavioral parameters. Running on mobiles that are low cost, user

[7]TCS Sustainability Report—Page 61; http://www.tcs.com/about/corp_responsibility/cs-report/Documents/GRI-2014-Sustainability-Report-071015.pdf.

[8]http://phx.corporate-ir.net/phoenix.zhtml?c=176060&p=irol-newsArticle&ID=2002794.

[9]http://thefinancialbrand.com/53498/digital-banking-lending-growth-opportnity/.

[10]**Mobility 2030:** Meeting the challenges to sustainability published by The World Business Council for Sustainable Development (WBCSD). Full report, Page 35 Link: http://www.wbcsd.org/Pages/EDocument/EDocumentDetails.aspx?ID=69.

friendly, and widely used, these applications can move healthcare from an illness driven system to a wellness driven system.

The quest for knowledge has been a great cause for travel. It made explorers undertake perilous voyages. It pushes us to explore the solar system and beyond. Digital technologies democratize exploration: many online maps depend on tens of thousands of 'people in the street' to map our land, sea, and even space.

In 2012, more than 4.5 million students were enrolled in tertiary education outside their country of citizenship. This number is only expected to grow.[11] Digital technologies enable education through Massive Open Online Courses (MOOC). The best universities around the world are offering online courses. This may make a minor dent in patterns of mobility.

Even if there are sustainability gains to be made by reducing some types of travel, the 'experience' seeking travel will continue to grow. Here too, digital can bring in some intervention. New and improved virtual reality products promise to provide virtual social spaces where people can participate in thrilling adventures or games together with a greater sense of real space and immersive movement. The FIFA Global Stadium, a digital sports platform, was used by over 1 billion fans from around the world reached the game: a number 13,000 times more than the stadium capacity.[12] It provided social, visual experiences, through a mobile app, that were more intimate and more immersive than TV.

Mobile devices are also changing the travel experience. A number of chores that required some amount of physical mobility no longer do so: shopping, banking, ticket booking, gifting and connecting with friends and family. And several travel related services are rendered by smart phones directly to consumers. Friends recommend holiday destinations on social networks; travel enablers such as flight availability, booking, logistics, directions, information on things to do and places to eat at are all available online, fueling the desire to travel. The smart phone has become a device where virtual and real travel converge.

6 Conclusion

Our world is going digital. Facebook with 1.05 billion daily active users[13] can be viewed as the third most populous entity after India and China. The mobile phone subscription count exceeds the world population count.[14] Much of the content that

[11]"Education at a Glance—2010" published by Organisation for Economic Co-Operation and Development.

[12]http://www.fifa.com/worldcup/news/y=2014/m=7/news=attendance-in-fifa-s-global-stadium-exceeds-1-billion-fans-2404624.html.

[13]http://newsroom.fb.com/company-info/.

[14]http://www.ericsson.com/res/docs/2015/mobility-report/ericsson-mobility-report-nov-2015.pdf.

flows through mobiles is processed in the cloud. To process these huge Big Data streams, analytics and Artificial Intelligence are required.

The digital consumer demands services online and on the mobile, 24 by 7 and in real time. Every sector is transforming itself to meet this demand. In the travel and transportation sector, computing power and digital technologies are creating continual improvements. They are also creating the big breakthroughs in products and services. Digital technologies are reimagining not just why we travel and how we travel, but the entire mobility ecosystem.

References

1. Arijit Chowdhury, Avik Ghose, T. Chakravarty and P. Balamuralidhar, "An Improved Fusion Algorithm for Estimating Speed from Smartphone's INS/GPS sensors", (Next Generation Sensors & Systems: Smart Sensors, Measurement & Instrumentation 16), (Ed:) S.C. Mukhopadhyay, [DOI: 10.1007/978-3-319-21671-3], Springer, Switzerland, pp 235–256.
2. T.M. Pollock and J. Allison, "Integrated Computational Materials Engineering: A Transformational Discipline for Improved Competitiveness and National Security," The National Academies Press, National Research Council, Washington, D.C., 2008.

Willingness to Pay for Improvement in Service Quality of Public Transit System

Vinod Vasudevan and Keshav Lunani

Abstract This paper focuses on the willingness to pay for different levels of improvement in travel time and comfort. The study clearly illustrates that passengers are willing to pay provided the service qualities are improved. The study also calculates the monetary value of increase in ratio of fare for specific levels of improvement in service qualities. These results could be useful for operators of the public transit system across India to introspect their policies aimed at improving ridership.

Keywords Willingness to pay · Transit users · Public transit systems

1 Introduction

Rapid growth in urban population has created extreme strain on all transport systems. The share of private vehicles is constantly increasing owing to reduction in the share of public transportation. This is the cause large part of congestion on roads and hence slowing down the speed. Tentative reasons for the shift could be due to liberalization of automobile industry, increase in the per capita income owing to increase in vehicle ownership, inferior quality of service provided by public bus service, discomfort relating to space and seat, unreliable schedule. Hence, to increase the share of public transport in developing countries there is a need to efficient transportation system which can carry high volume with better comfort and in reasonable time through congested roads of urban areas. This requires increased investment in public transit systems. The transit authorities, especially those in developed countries, face fund crunch and hence the investment has to come from the users.

V. Vasudevan (✉) · K. Lunani
Department of Civil Engineering, Indian Institute of Technology Kanpur, Kanpur, India
e-mail: vinodv@iitk.ac.in

K. Lunani
e-mail: lunanikeshav@gmail.com

M.C. Tandon and P. Ghosh (eds.), *Mobility Engineering*,
DOI 10.1007/978-981-10-3099-4_9

2 Objective

The objective of this paper is to investigate passengers' willingness to pay for improvement in service quality of Indian buses and determine factors influencing the value passengers' bid for different improvement in travel attributes. The knowledge of willingness to pay can be used to suggest improvement in the urban bus service and hence maximizing the user benefits. This study will help to develop economic strategies for improving bus service in India and reduce vehicle density on roads hence improving the environment conditions of urban areas.

3 Data Collection

Stated preference surveys were conducted in three cities: Chennai, Hyderabad, and New Delhi over three months. A total of 937 responses were collected from these cities.

4 Data Analysis

Collected data from the questionnaire survey included the willingness to pay for improvement in travel time and for improvement in comfort. Since definition of comfort is very subjective, graphics were shown for illustration. Figure 1 shows the summary of WTP for improvement in travel time based on data from New Delhi and Fig. 2 shows the WTP for improvement in comfort based on data from New Delhi. The data from Chennai and Hyderabad also showed similar trend. These data shows that, on an average, people are willing to pay only for higher improvements of travel time. On the other hand for improvements in comforts, people are willing to pay even for smaller levels of improvements.

Fig. 1 WTP for improvement in travel time in Delhi

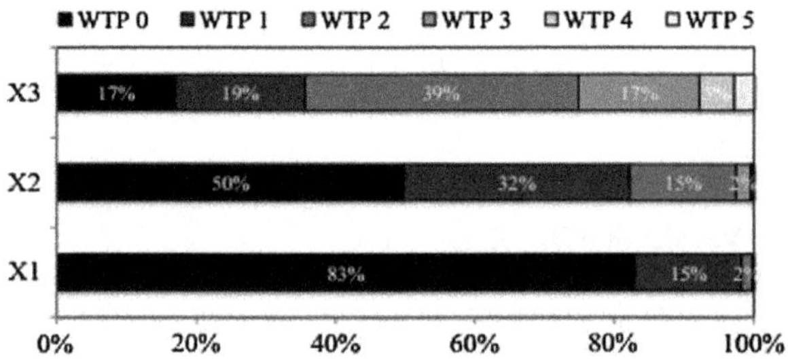

Fig. 2 WTP for improvement in comfort in Delhi

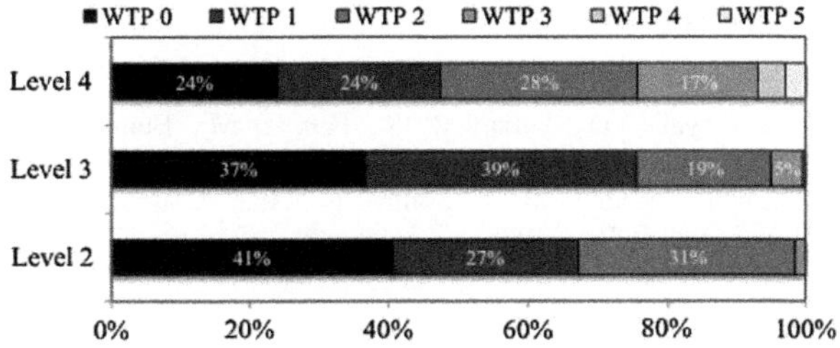

Mean willingness to pay values were estimated for both these cases using equation proposed by Dissanayake et al. [1].

$$W = \sum_{n=1}^{N} \frac{\left[\sum_{\forall i} P_n(i) w(i)\right]}{N}$$

where,

$P_n(i)$ Probability that individual n chooses alternative i

W(i) Willingness to pay choice option i expressed as ratio of choice in fare to their present fare

N Total number of individuals

W Mean willingness to pay in terms of ratio of maximum fare

5 Results

Mean willingness to pay values for improvement of travel time by one scale ($\sim$1/6th of the current travel time) was 1.08 when data from cities are considered. The same value for improvement in comfort is 1.21, indicating that people in the surveyed urban areas are more concerned about the comfort than the travel time.

6 Conclusions

The findings from this study shows that people in urban areas are willing to pay for improved quality of service. Although it is difficult to compare levels of comfort and travel time, the results indicate that people's willingness to pay is higher for improved comfort than improved travel time.

Reference

1. Dissanayake, D., Umar, R. S., Faudzi, M., Bentotage, S., Widyastuti, H., & Herijanto, "Inter-regional Valuation of Road Causalities and Exploration of Social Attributes to Road Casuality Reduction in Southeast Asian Countries, Thailand, Malaysia and indonesia. Indonesia" ICRA (2008).

Shinkansen/High Speed Railway

Harutoshi Hayasaka

Abstract My presentation consists of JIC's Mission, HSR Projects JIC in hand, Shinkansen Network in Japan, Comparison of Seikan Tunnel and Channel Tunnel, the Longest Railway Tunnels in the World and the like.

Keywords Shinkansen · HSR · JIC · Seikan tunnel · Channel tunnel · SENS

1 Brief About the Speaker's Major Jobs

I'm honored to be invited to deliver the keynote presentation in Session III: Modern Rail Transport Systems of CAETS Symposium 2015: "Mobility" Engineering of the Indian National Academy of Engineering (INAE). From now on, I'd like to introduce my major works. When I was 20's, I was in charge of construction of the Seikan Undersea Tunnel in Japan. Next, when I was 30's, I was in charge of supervision of the Channel Tunnel construction as one of Technical Advisors to the Banking Syndicate for the Eurotunnel Project Financing. Then, 3 years ago, I was in charge of Pre-F/S for HSR Line-6 between Chennai and Trivandrum with branch line to Mysore in India as the Chief Railway Expert cum team Leader. Recently, I'm now in charge of Engineering Service for making Standard Technical Specification of Swedish High Speed Railway. And, I'm in charge of Joint F/S for Mumbai-Ahmadabad, High Speed Railway Corridor in India as Acting Leader.

2 Our Mission—Project Stage

JIC is able to manage every project stage and field.

JIC's railway project operations extend from the upstream to downstream stages.

H. Hayasaka (✉)
Japan International Consultants for Transportation Co., Ltd., Tokyo, Japan
e-mail: hayasaka@jictransport.co.jp

M.C. Tandon and P. Ghosh (eds.), *Mobility Engineering*,
DOI 10.1007/978-981-10-3099-4_10

In many cases, as you know, Railway Project starts from the Project Preparation Stage.

And then, we will go to the Engineering Service Stage in which Basic Design and Detailed Design shall be done.

And, Tendering Assistance and Negotiation, Supervision of Construction Works, and at the final construction stage, Various Verification and Commissioning Test are carried out.

3 Company Profile Our Assets and Capital

As to the Asset Information, our capital amounts to 495 million yen (approximately 5 million dollars) As for our capital contribution ratio, JIC is 53% owned by JR East, and thus JIC is subsidiary in terms of JR East consolidated accounting, marketing JR East our parent company.

4 Project Map

Countries;

(1) JARTS implemented studies till March, 2012
(2) JIC is involved since April, 2012.

Map shows projects implemented by both JARTS and JIC from 1965 and 2012. The yellow is from 1965 onward, while the red indicates projects after JIC full-fledged operation in 2012. As can be seen in this map, the projects are being conducted worldwide.

5 HSR Projects JIC in Hand

In Sweden, Engineering Service on high Speed Railway is now carrying out from Sept. 2014 for 2 years for formulating Technical Standard of Swedish HSR.

6 Outline of High Speed Rail Projects Since Fiscal 2000

In the year 2012, JIC conducted some Pre-F/S on HSR in India. Namely, HSR No. 5 and No. 6 corridors. The author was a Chief Railway Expert cum Team Leader of No. 6 HSR Pre-F/S in India.

7 Shinkansen Network in Japan

Japan's Shinkansen Network has more than 2600 km in length. In spring of 2016, a section between Shin-Aomori and Shin-Hakodate/Hokuto will be opened.

8 Main Features of the Shinkansen

The main features are safety, punctuality of train operation, high density operation mass transit.

9 Early Earthquake Detecting System

Shinkansen system has the early earthquake detecting system which consists of seismometer at coast, substation and seismometer along a railway line.

10 Annual Average Delay Time in Tokaido Shinkansen

The minimum headway of Tokaido Shinkansen is 3 min. Total 336 trains are operated per day, up and down directions. Under such high-density operations, an annual average delay time in Tokaido Shinkansen is only 0.6 min in FY 2010.

11 Frame Work for Implementation of Shinkansen Project

National Government bears financing burden for the Shinkansen infrastructure with two thirds, while local governments bear those with one thirds.

12 Procedure of Shinkansen Construction

National Government, Ministry of Land, Infrastructure, Transport and Tourism (MLIT) decides a basic plan, then, direct and order to Japan Railway Construction, Transport and Technology Agency (JRTT) for conducting study. Afterwards, the construction begins and completed. JRTT leases the railway infrastructures to JR as railway operator.

13 Structures on Shinkansen in Japan

Main structures are tunnel, bridge, viaduct and earth structure. Around 70% of the Japan proper composes mountainous areas. As a result, there are many tunnels of the Shinkansen lines.

14 Comparison of Seikan Tunnel and Channel Tunnel

The Seikan Tunnel commenced a commercial train operation on Mar. 1988, then, became a commercial artery between Honshu and Hokkaido in Japan. The Channel Tunnel also commenced a commercial operation on Nov. 1994, and became the commercial artery between UK and France. Both tunnels have more than 50 km long length.

15 Location of Seikan Tunnel

The Seikan Tunnel is the undersea tunnel in Japan.

16 Geological Profile of Seikan Tunnel

Geology of the Seikan Tunnel consists of rhyolite, sandy or silty mud stones and tuff breccia and the like.

17 Geological Profile of the Channel Tunnel

The geology of the Channel Tunnel consists mainly of Mesozoic cretaceous layer. This layer is by much older than the Neocene igneous rocks and sedimentary rocks of the Seikan Tunnel.

18 Cross Section of the Seikan Tunnel

The Seikan Tunnel consists of three tunnels such as the main tunnel, the Service Tunnel and the Pilot Tunnel.

19 Percentage of Structures on Shinkansen

The more becomes newer era, the more increases percentage of concrete structures rather than earth structures.

20 The Longest Railway Tunnels in the World

Among 15 tunnels, the author was in charge of four tunnels in engineer's life. These are the Seikan Tunnel, the Channel Tunnel, the Hakkoda Tunnel and the Iwate Ichinohe Tunnel.

21 A New Construction System "SENS"

The SENS has been developed to create a new reasonable method for projects classified in the boundary range between the mountain tunneling (NATM) and the shield tunneling.

22 Comparison of Surface Settlement Between SENS and NATM

To date the SENS has achieved an advance rate double to triple that of the NATM in excavation of the Shinkansen double-track tunnels in unconsolidated ground.

References

1. TUNNELLING ACTIVITIES IN JAPAN 2012, JTA, (2012)
2. TUNNELLING ACTIVITIES IN JAPAN 2014, JTA, (2014)

Unique Engineering Features of Delhi Metro

Mangu Singh

Abstract When planning of construction of metro system in Delhi was started in 1997, there was no experience of constructing modern metro in the country. Only experience was constructing Kolkata Metro North-South Line which took 24 years to complete 16.5 km Line. Therefore, many innovating engineering solutions were adopted in developing the metro system in Delhi to ensure speedy construction and safety of public and property during construction with least negative impact on environment. Further in order to have financial sustainability of the system, energy efficient, cost effective solutions were deployed.

Keywords Urban rail transit · Metro rail system · Underground railway · Tunnelling in urban areas · Segmental construction

1 Introduction

Though the need for a rail based public transport system for the city of Delhi was felt a long back but it is only mid 90's when Phase I of Delhi Metro project was sanctioned by the Government. The responsibility of construction of rail based mass transport system was entrusted to Delhi Metro Rail Corporation Ltd., a Special Purpose Vehicle, formed for this purpose. The Phase-I of the project was sanctioned by the Government in Year 1995, however, the actual planning and execution of work were started only in 1997 when Delhi Metro Rail Corporation came into real existence. Phase-I of the project consisting of 3 Line, totalling of about 69 km, was completed in 2005 i.e. in about 7 years 3 months time, 2 year 9 months before the original programme of revenue opening date. The first Section of small length of about 8 km, however was opened in December 2002.

Phase II of the project consisting of about 125 km was mostly completed by October 2010, before Common Wealth Games held in October 2010 in Delhi.

M. Singh (✉)
Delhi Metro Rail Corporation (DMRC), New Delhi, India
e-mail: mdmetro@hotmail.com

M.C. Tandon and P. Ghosh (eds.), *Mobility Engineering*,
DOI 10.1007/978-981-10-3099-4_11

Phase II was also completed within time and cost. The included Airport Express Line of about 22.5 km executed on PPP basis. After completion of Phase-II, a network of about 190 km became operational.

Phase-III of about 140 km is under implementation and is likely to be completed by December 2016. Three small Section of Phase III, totaling about 22 km are already made Operational.

When planning of construction of Metro System in Delhi was started in 1997, the biggest challenge was to decide on the system parameters. At that time the experience of modern Metro system within the country was very limited. A Line of 16.45 km was only operational in city of Kolkata which was constructed between 1974 and 1996. This Line was constructed with vintage type of technology available at that time. Therefore, this was not of any help for deciding the technology system for proposed Delhi Metro.

Delhi have peculiar characteristic of very heavy traffic demand on the selected Corridor. Delhi also have difficult climatic condition ranging from extreme hot, dry and dusty summer to very cold winters and storming raining seasons. Internationally also by that time, not many Metros were having experience of similar climatic and traffic conditions. Therefore, while deciding on technical parameters and systems, these issues were required to be taken into account. Further city roads being very congested could not have afforded for occupation for construction of Metro for long duration. City also did not have any experience of tunnelling in a very difficult and variable geology ranging from soft silty clay to weathered fractured rock to very hard quartzite rock. The construction methodology, technology and procedures were therefore, required to be selected to suit the requirement of speedy construction.

2 Construction of Underground Tunnels

Tunnels for Metro System are typically shallow with minimum over burden as deeper tunnels will necessitate deeper stations, inconvenient for Metro users and also energy inefficient station for deeper escalators and lifts. Shallow tunnelling in urban area poses difficulties in settlement control at surface to protect structures properties, life, traffic and utilities. In Delhi specially designed Shield Tunnel Boring Machines with tailor made specifications, to suit the geo-technical conditions have been deployed. The Condition Survey of structures in the influence zone of tunnelling carried out and effect of projected settlement on the buildings/ structures is examined in advance to ensure that the tunnelling is done safely without causing damage to the structures. Suitable modifications to the design are done based on the pre-condition survey and settlement analysis.

Pre-cast concrete segmental lining is used for tunnelling. The segment casting yards are located away from the busy central urban area. The segments are transported during night hours in large trailers to avoid traffic congestion during day time/peak hours. Similarly, batching plants for concreting works were also located outside the main urban area.

Since at most of the locations tunnelling is below ground water table, hydrophilic gasket is used to prevent leakage from the segment joints. A large number of tunnel boring machines were deployed so that completion of tunnels matches with the completion of station structures and optimum project completion period is achieved. With this in Delhi, the project completion of underground Section in 4½–5 years time from starting the bid process to finally opening of the line after necessary trials is achieved.

3 Station Construction

The underground stations are mostly located below major roads and, therefore, occupation of roads for construction of stations becomes very critical. Occupation of roads for longer duration causes inconvenience to the public as traffic is required to be diverted partially or fully. Therefore, top down method of construction was adopted instead of cut and cover construction method.

RCC diaphragm walls are first constructed with highest level of quality. This was possible due to advancement in diaphragm wall construction technology. Verticality of diaphragm wall panels was ensured by suitable trenching method and confirming trench verticality before reinforcement is lowered and concreting is done. The verticality is checked by recording the trench by special equipment like "Coden" and in case the trenching is not being vertical to the expected level, the trench is refilled and re-trenching is done. In order to prevent collapse of soil into the trench, special polymer slurry is used instead of bentonite to avoid the problem of disposal of environmentally hazardous benotonite slurry. Similarly, it has been possible to provide PVC water stops between the two diaphragm wall panels to prevent the leakage from the diaphragm wall joints. Once the diaphragm wall is completed along the perimeter of the station, the top slab of the station box is first completed in part or in one go as the case may be, leaving opening in the top slab for further excavation and construction of the structures below the top slab. With this, it is possible to restore the roads for public quickly and reduce the road occupation drastically.

4 Elevated Viaduct

The use of segmental construction in elevated viaduct/bridges was not common in India till DMRC started metro construction in Delhi. The construction of elevated viaducts in city areas if done with conventional cast-in situ method, will cause great inconvenience as this will need occupation of roads. Further it will take much longer duration to complete the viaduct works. DMRC, therefore, decide to go for precast segmental construction in big way.

The other advantages of precast segmental construction are:

Segmental construction is a cost effective method for a vast range of span lengths and types of structures. Structures with sharp curves and variable super elevation can be adjusted to suit the requirements.

Segmental construction allows the reduction of construction period as segments are manufactured and assembled rapidly thereafter whilst substructure work is in progress.

Segmental construction shields the environment as minimum space is required for foundation and sub-structure at site. The superstructure is manufactured at a place away from busy locations. Placement of superstructure is completed with the system erected from piers at heights.

In case the site situation demands then it is easy to adjust last minute changes in span configuration.

It is easier to ship smaller segments by road trailers on the city roads.

During construction, the intervention to the road traffic is considerably reduced.

Segmental construction results in aesthetically pleasing structures and good finishes.

Significant improvement in quality is possible in the casting yard.

During construction, the technique exhibits high record of safety.

5 Use of Precast Full Span U Girders

Use of precast full U girders not only speeded up the construction but also improved the aesthetics of the viaduct because of sleek shape and uniform quality.

Precast pier caps and pre-cast full span U girders are suitable for constructing long elevated viaducts for railway system.

Key advantages of the system

- Integrated sound barrier.
- Integrated cable support and system functions.
- The longitudinal profile of Track could be lowered by 1.5–2 m compared to regular design.
- Integration of the walkway for upkeep and passenger evacuation on each side of the tracks.
- Cost effective from both structural and system point of view.
- Flexibility of construction starting points along the line.

The viaduct viz; Airport Metro Express Line, Faridabad Metro Line, Noida-Greater Noida Metro Rail Line and Kochi/Lucknow Metro Rail Line uses full–length double precast U girders.

6 Use of Special Structural Forms like Extradosed Bridge

Extra-dosed bridges are the best structural solution in places where long spans are required to be constructed and there is space constraint. They are also more economical than the cable stayed bridges. The pylon heights in extradosed bridge are almost half in height as compared to cable stayed bridges.

For the extension of the U shaped viaduct of line-3 of Delhi Metro, DMRC dealt with the crossing of five railway tracks with numbers of constraints, sharp plan curvature, railway vertical clearance, impossible location of intermediate piers and unavoidable interruption of the railway traffic. Besides technical difficulties, the line opening date was scheduled within less than one year.

The Pragati Maidan bridge is an extradosed railway bridge with a main span of 93 m. The deck cross-section has a U shape, which allows a great integration of the metro system in the supserstructure. The extradosed cables are wrapped by a concrete beam that allows to regard them as internal prestressing. This beam also enhances the stiffness of the main span.

The extradosed bridge involved a very complicated construction technique as it had to be constructed over five very busy lines of the Indian Railways. With almost 200 trains, comprising many key trains going towards western and eastern India, passing on the tracks under the new bridge day in and day out, it had to be constructed without any interruption to the train traffic. Besides first bridge in Japan, this is second bridge of its kind built over a railway crossing.

It is the first extradosed bridge ever built in India to be erected using the cantilever construction technique.

7 Selection of Rolling Stock

For a public transport system like metro, financial sustainability of the system depends on keeping the cost of operation and maintenance low. The cost of energy normally constitute around 30–40% of the operational const. It is, therefore, necessary to have very energy efficient System.

DMRC has successfully deployed highly energy efficient metro rolling stock with very high reliability, maintainability and safety standards. Stainless steel car body has been chosen to achieve light weight & less maintenance on account of corrosion. All materials used in the stock are of very high fire rating. Propulsion equipment design has been chosen to achieve high acceleration coupled with highest possible regeneration of energy during braking. Passenger comfort has been kept in focus and the design incorporates two level suspension system with superior bogie design for improved ride quality. State of the art passenger announcement systems and passenger information system coupled with dynamic route map provide requisite information to commuters. High capacity energy efficient on board air conditioners have been provided. Car body construction ensures excellent thermal and acoustic insulation to achieve interior conditions consistent with the requisite international standards. To improve the rail wheel interaction and for mitigation of noise and wheel wear, wheel flange lubricators are used with bogie design suitable for negotiation sharp curves. Braking characteristics are designed for full regenerative braking over almost the complete range of operation and apart from improvement in energy consumption: this would help in reduction in friction braking requirements. DMRC is committed to achieve maximum indigenization and we have achieved notable success and almost 90% of the stock deployed in DMRC has been manufactured in India. By adopting regenerating type of rolling stock technology, DMRC become first organization in the world to have registered CDM Project with UNFCC in rail based transport sector and has earned carbon credit.

8 *25 kV* AC Traction System

Till then most of the metro systems in the world have DC traction system, mostly around 750 V with 3rd rail current collection system. The low voltage DC system has limitations on the capacity of the system and, therefore, some metros started using 1500 V DC system with overhead catenary. The high voltage is necessary for higher capacity as more numbers of trains are required between 2 Sub-Stations drawing very high current, Delhi Metro, since required to have very high capacity ranging from 60,000 passengers per hour per direction to last capacity of 90,000 passengers per hour per direction, explored the possibility of further high voltage and finally selected 25 kV AC traction system which is very common for main line operation throughout the world. Therefore, for the first time for metro type of operation, 25 kV AC traction system was adopted in Delhi Metro and since then the system is working very satisfactorily. The light weight efficient rolling coupled with high voltage AC traction system has proved to be very energy efficient system and helped Delhi Metro Rail Corporation in maintaining its financial viability.

9 Automatic Fare Collection System

It was also realized that having an efficient metro network necessarily required a passenger friendly, state-of-the-art fare collection system.

The Automatic Fare Collection System has lesser requirement of staff, lesser possibility of leakage, amenable to quick fare changes, all transactions are logged and management information reports easily generated. Hence the obvious choice between the manual and the automatic system was the Automatic Fare Collection System (AFC).

Again the choice was between a magnetic recyclable ticket/contact media or a contactless media. The analysis done by DMRC lead to the conclusion that contactless solution is the optimum solution as it provides better through put, more number of recycling of the ticketing media and less prone to mechanical problem due to contactless interaction between the validator and the ticket. Contactless smart media cards were considered ideal for multi journey but the single journey dilemma remained on cost considerations. Finally, CST's were found to be the right choice for single journeys. The single journey token themselves are initially more expensive to produce, but they are recovered at gate exits and can be reused up to 50,000 times, reducing the unit cost per journey.

DMRC was first metro system in the world to introduce this type of contactless smart token for single journeys. Today, however many metro systems in the world have adopted this system.

10 Conclusion

The number of unique engineering features adopted in Delhi Metro has helped in developing metro system at very rapid pace to provide relief to commuters in Delhi NCR from severe congestion on roads. DMRC is providing safe, comfortable and very reliable transport system to city of Delhi and NCR. The trains are running with more than 99.7% punctuality. With the application of energy efficient technology, the economical benefit from the project are estimated in the range of 20–25%. The system has also helped in reducing the carbon-di-oxide emission and also reduction in pollution level in Delhi and NCR.

Creating Sustainable Transportation System in India—Materials Engineer Perspective

Aravind Krishna Swamy and Priyansh Singh

Abstract Construction and maintenance of road infrastructure requires large amount raw materials, manpower and energy. Increased environmental awareness, shortage of raw materials and legal obligations are prompting engineers to look/adopt for sustainable pavement design and construction technologies. This paper discusses several sustainable practices that can be used by transportation engineering community including waste materials, alternate materials, better material behavior modeling techniques, and construction practices.

Keywords Sustainable construction · Pavement material · Recycling · Continuum damage · Pervious pavement

1 Introduction

India has a second largest road network across the world with total length of 4,689,842 km [1]. Even tough this translates to 0.66 km/square km, some areas within the country lacks proper connectivity. This lack of connectivity coupled with economic growth has prompted government to invest heavily on developing/upgrading transportation infrastructure including (i) road network (national highways to rural road), (ii) urban infrastructure (elevated roads, interchanges), and (iii) airports. All these infrastructure development activities require huge quantities of construction materials. This includes naturally available materials (Eg. soil, aggregates) as well as manufactured materials (cement, steel). However, with depleting natural resources, cost of these high quality materials is increasing with time. In some parts of the country, such materials are available in scarce. This scenario is forcing material engineers to look for cost effective, and less polluting construction materials/practices as well as pavement performance prediction

A.K. Swamy (✉) · P. Singh
Department of Civil Engineering, Indian Institute of Technology Delhi,
New Delhi 110016, India
e-mail: akswamy@civil.iitd.ac.in

M.C. Tandon and P. Ghosh (eds.), *Mobility Engineering*,
DOI 10.1007/978-981-10-3099-4_12

methodologies. Some of the pavement design techniques/materials/practices that can lead to sustainable environment are discussed below.

2 Use of Recycled and Waste Materials

The selection of raw materials is a vital phase of any civil engineering construction. The good selection practice can significantly decrease the initial cost and as well as maintenance cost. This selection of materials is often governed by availability, transportation to site, compatibility with other materials, and long-term performance. This issue can be addressed to certain extent by recycling the materials from the existing roads and by using the alternative materials.

Reclaimed Asphalt Pavement (RAP): It is the construction material removed from deteriorated roads and processed so as to reuse in new construction. Under suitable circumstances, aged asphalt binder present in RAP mixes well with the virgin binder and new aggregates. Depending on constituent material properties, mixing technology, the properties of recycled mixture can be as good as fresh mixture and can be utilized for the construction of new pavement. Due to higher stiffness of RAP binder, recycled asphalt concrete mixturesoften leads to better performing mixtures under rutting and poorly performing mixture under fatigue [2, 3].

Recycled Concrete Aggregate (RCA): RCA is processed from the old concrete structures. It's most popular stone replacement for construction purpose. RCA can be easily used in the place of crushed stone in the base, granular and surface courses of Highway. Due to increased porosity, cement concrete containing RCA requires more cement paste when compared to conventional concrete mixture. RCA is collected from different sources; homogeneity is often a major issue in using it in pavement construction.

Industrial Waste/Byproducts: Fly ash, silica fume, slag, foundry sand, kiln dust and other byproducts which has good mineral content, pozzolonic properties that can be easily used as a partial replacement of soil/aggregates or as an admixture for asphalt and concrete mixtures. These materials can be used in bulk fill areal like embankments, retaining walls etc.

Plastic, and Rubber: Tires and other domestic plastic waste contain certain elastomers/plastomers, which are compatible with asphalt. Thus, these materials can act as asphalt binder modifier. The modified asphalt mixture has a higher stability, reduced pavement deformation; increase fatigue resistance and provide better adhesion between the asphalt and the aggregate [4].

3 Mechanistic Characterization

Due to recent advancements in the field of material modeling the more efficient techniques are available. The viscoelastic continuum damage approach relates stress-strain through thermodynamic principles. Thus, this approach requires lesser number of specimens, testing time and resources [5, 6].

4 Construction Practice

Cold Mix Asphalt: Cold mix is a mixture of unheated aggregate and emulsion/cutback and filler. In case of cold mix asphalt mixture, emulsion (or cutbacks) are mixed with aggregates at ambient temperature (10–30 °C). Thus cold mixtures eliminates heating requirement and thus conserves energy. Cold mix pavement can provide energy savings of over 50% compared with hot mix [7]. Open graded mixtures are storable and semi dense mixtures have good adhesion and lower permeability.

Warm Mix Asphalt (WMA): In case of WMA, viscosity of binder is reduced by addition of certain waxes (example sasobit) or foaming action. This binder with reduced viscosity is mixed with aggregate at 100–140 °C. Reduced mixing and paving temperatures decreases the energy needed to produce asphalt concrete, reduces emissions and odors from plants, and provide better working environment at the plant and paving site [8].

Permeable Pavement: This type of pavement consists of various layers i.e. base, subbase etc. that allow the movement of water through them. Thus decreasing surface runoff while recharging ground water. The permeable pavement can be used as a supportive drainage system for urban areas. However these pavements requires frequent maintenance due to clogging of pores [9].

5 Conclusion

The implementation of these sustainable pavement design/material/construction technology in developing countries is challenging. Adoption of new technologies and usage of waste materials would incur additional cost for the constructor. Thus certain incentives have to be provided from government/implementing agencies. Also, new technologies/construction practices require better understanding with respect to conditions. Therefore long-term field studies have to be conducted which lacks at this point of time.

References

1. Indian Road network, (https://en.wikipedia.org/wiki/Indian_road_network) last retrieved on 1/October/2015) (2015).
2. Kandhal, P.S, Rao, S.S., Watson, D.E., Young, B. Performance of recycled hot mix asphalt mixtures. NCAT Report No. 95-1. National Center for Asphalt Technology, Auburn University, Alabama, (1995).
3. Aravind, K., and Das, A. Pavement Design with Central Plant Hot-mix Recycled Asphalt Mixes, *Construction & Building Materials*, 21(5), 928–936, (2007).
4. Awwad, M. T. and Shbeeb, L., "The use of polyethylene in hot asphalt mixtures," American Journal of Applied Sciences, 4(6), 390–396 (2007).
5. Bhattacharjee, S., Swamy, A.K., and Daniel, J. S. "Application of the Elastic-Viscoelastic Correspondence Principle to Determine the Fatigue Endurance Limit of Hot Mix Asphalt." *Transportation Research Record*. 2126(1), 12–18, (2009).
6. Swamy, A.K.. Evaluating Mode of Loading Effect and Laboratory Fatigue Performance of Asphalt Concrete Using Viscoelastic Continuum Damage Mechanics. Doctoral dissertation, University of New Hampshire (2011).
7. Doyle, T., "Relating laboratory conditioning temperature to in-situ strength gain for cold mix pavement in Ireland," Ph.D. thesis, University College Dublin, Dublin, Ireland, University College Dublin, Dublin, Ireland.
8. You, Z. and Goh, S. W., "Laboratory evaluation of warm mix asphalt: A preliminary study," International Journal of Pavement Research and Technology, 1(1), 34–40 (2008).
9. Scholz, Miklas, and Piotr Grabowiecki, "Review of permeable pavement systems," Building and Environment, 42.11 (2007).

Lack of Enforcement Nullifies the Benefits of Mobility Improvement Measures: Some Examples

L.R. Kadiyali

Abstract Based on a study of measures adopted for increasing the safety and mobility of road users, it is found in many cases that the safety measures built at great cost are misused. Lack of enforcement nullifies the benefits which were expected from the improvements. Several examples from different parts of the country on National Highways and Arterial Streets will be presented. The presentation will highlight the need for strict enforcement of traffic rules where engineers safety measures have been proposed and are built at great cost.

Keywords Traffic regulation enforcement · Urban mobility · Urban transportation · Pedestrian traffic · City traffic management

Under the large programme of road improvements, India is investing considerable amount of money. Various schemes of road widening, strengthening of old pavements, construction of grade-separated interchanges, pedestrian facilities, separate cycle tracks, bypasses, etc. are being implemented. A study of a few completed schemes by the Author has revealed that lack of proper enforcement measures has nullified the benefits intended to accrue to the society. The Author has selected two such schemes to illustrate how the money invested on the improvement measures has not served the objectives of the schemes. The schemes selected are:

(i) Mukarba Chowk Interchange in Delhi
(ii) Pedestrian subway in Tumkur on NH 4 in Karnataka

L.R. Kadiyali (✉)
New Delhi, India
e-mail: lrkadiyali@gmail.com

M.C. Tandon and P. Ghosh (eds.), *Mobility Engineering*,
DOI 10.1007/978-981-10-3099-4_13

1 Mukarba Chowk Interchange in Delhi

Till recently (up to 2007), the crossing between the Ring Road and Azadpur GT Karnal Road (NH 1) was at ground level. The junction was signalised. As there is a big market (Mandi) at Azadpur, truck traffic on the Azadpur GT Karnal Road was heavy and unmanageable. Similarly, the Ring Road which caters to large bus traffic and tuck taffic, was having a very high volume of traffic. The at-grade signalised junction was a major traffic bottleneck unable to handle the traffic in the $4 \times 3 = 12$ directions. The Govt. of NCT Delhi thought of implementing a grade separated structure. After calling for competitive quotations, the PWD selected the consortium of Tandon Consultants, L.R. Kadiyali and Associates and Archohm to do the detailed design. The Consultants came up with the design of a cloverleaf interchange which catered to all the movements. Separate provision was made for bus-bays, cycle tracks and pedestrians. The design incorporated the best available standards, keeping in view the safety of all road users. The layout of the interchange is given in Fig. 1.

1.1 Review of Present Traffic Situation

Several visits were made by the Author to review the present traffic situation in the various locations, with a view to understand whether the traffic movements conform to what was envisaged in the design. It has been noticed that the traffic pattern now present, grossly misuse the facility thus resulting in compromising safety. These are discussed in detail below.

1.1.1 Misuse of Cycle Tracks

The original design had provided separate cycle tracks to facilitate the free movement of cyclists (See Fig. 2). During the visit to the site on the 31st August, 2015, it was noticed that cycle tracks were used by two-wheelers, three-wheelers and small cars. This is highly undesirable. The entry to the cycle tracks should be provided with steel bollards spaced so closely as to admit only cyclists. Alternatively, security personnel may be deployed to physically prevent misuse of cycle-tracks (Photos 1 and 2).

To prevent motorised vehicles from using the cycle track, it may be necessary to

(i) plant iron bollards spaced suitably such that only cycles can enter.
(ii) post security guards to enforce the traffic discipline.

Fig. 1 Layout of grade-separated inter-change at Mukarba Chowk

1.1.2 Misuse of Pedestrian Concourse

The design proposed by the Consultant had proposed all pedestrian movements to take place at a level (Level Minus 1) below the present Azadpur-Karnal Road which is at ground level (Level 0). In order to facilitate the pedestrians to reach the concourse level which is below the ground level, escalators and ramps were provided. It was envisaged that at the pedestrian concourse, there will be no vehicular traffic and no cycle traffic, and pedestrians from any of the four arms of the Mukarba fly-over could easily reach the subway concourse and reach any arm of their desired destinatioin at the ground level (Level 0) or flyover level (Level Plus 1).

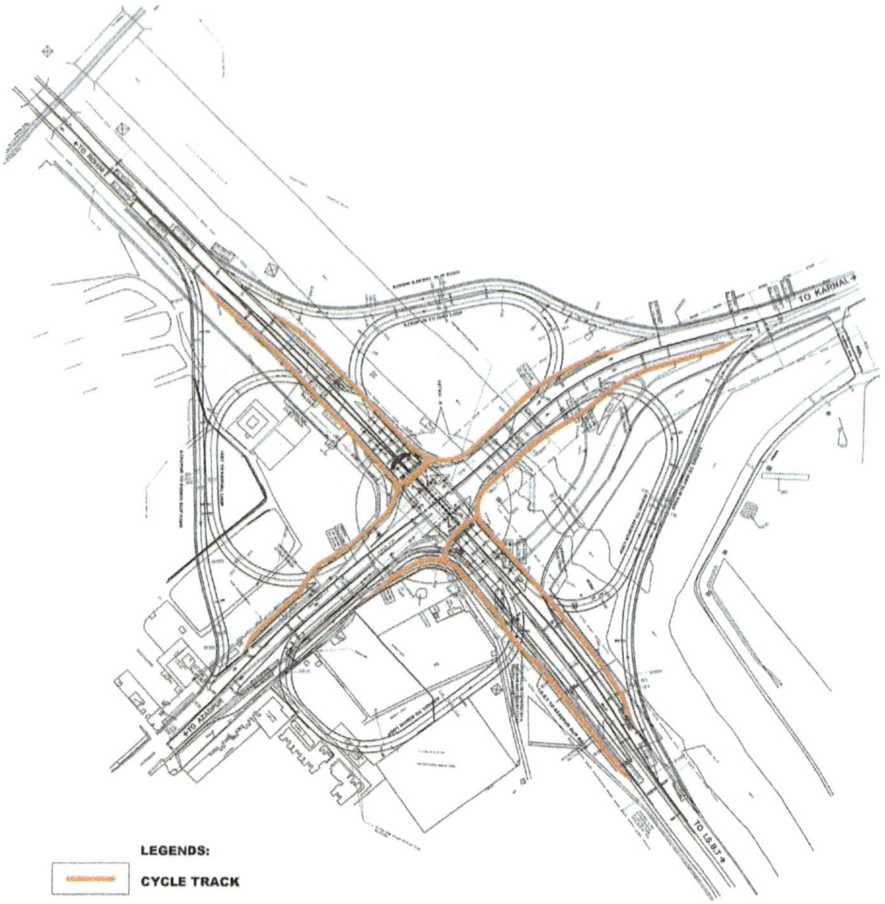

Fig. 2 Layout of cycle track at Mukarba Chowk

Level Plus 1 (Flyover)
Level 0 (Ground Level)
Level Min 1 (Pedestrian Concourse)

During inspection of the site, it was found that the pedestrian concourse was being used by cyclists, two-wheelers, three-wheelers, cars and vans (Photo 3). This is a gross misuse of the pedestrian facility, endangering the safety of pedestrians. The facility being below ground level, there is a need for adequate lighting. Only one lamp was functioning. Some of the escalators were also not in working order.

Photo 1 Cycle track being used by an autorickshaw

Photo 2 Four-wheelers using cycle tracks

The following short-term measures are proposed to prevent the misuse of the pedestrian facility and to improve its usefulness:

1. All entry points should be provided with iron bollards thus preventing the entry of cyclists, two-wheelers, three-wheelers, cars and vans.
2. The non-functioning escalators should be immediately repaired.
3. The illumination of the pedestrian concourse should be improved so as to have a lux of at least 20.

Photo 3 Four wheelers moving inside pedestrian sub way

4. The large unused space in the pedestrian concourse can be rented out to shoppers. This will not only earn revenue but also increase the safety of pedestrians, particularly women.

2 Pedestrian Subway in Tumkur on NH-4 in Karnataka

On a busy section of the NH-4 passing through Tumkur town, there are educational institutions and hospitals on either side. Students cross the road subjecting themselves to great danger of accidents (Photo 4). To remedy the situation, a pedestrian

Photo 4 Girls crossing the road

Photo 5 Locked subway entrance

subway has been built. During site visit by the Author, it was seen that the entrance to the subway was locked (Photo 5).

Discussions held with the local people revealed that the subway was not lighted properly, and was being misused by beggars and the homeless and was not safe for women to use. Hence it was closed. This clearly shows that the proper enforcement measures like posting a security guard, cleaning the premises regularly and providing good lighting, the subway would have been safe for use. In the absence of such enforcement measures the costly investment on the construction is rendered infructuous.

3 Conclusion

The two examples show clearly that along with the provision of improvement measures, adequate enforcement measures should be put in place to ensure that the facility is being used as intended.

Optimizing Traffic Control for a Minimization of Fuel Consumptions and Emission Values

Zsuzsanna Bede and Péter Gáspár

Abstract The emissions of harmful substances to the environment can be reduced in two ways. On one hand, by using the Adaptive Cruise Control (ACC) system of individual vehicles, on the other hand, by controlling the vehicles with the help of appropriate traffic control tools. These two approaches can be analysed in a complex model consisting of the models describing the behavior of the individual vehicles and determining the operation of the traffic network and the relationships between them.

Keywords Adaptive cruise control · Emission-based traffic control

1 Introduction

The aim of the ACC is to adjust the individual speed of a vehicle, by taking into account the topographical characteristics of the road, the speed limit and the current traffic conditions [1, 2]. The traffic data covers the Signal Phasing and Timing (SPaT) information related to traffic light and the regulatory, warning, and guidance information displayed by Variable Message Signs (VMS). Knowledge of the local data regarding the environment of a vehicle or the global data obtained from the traffic control center are required for Vehicle-to-Vehicle (V2V) and Vehicle-to-Infrastructure (V2I) communications. The vehicles are able to receive information by means of an electronic on-board unit (GPS, TMC, FCD etc.), and thus theirs routes can be optimized according to some criterion (shortest road,

Z. Bede (✉)
Department of Control for Transportation and Vehicle Systems,
Budapest University of Technology and Economics, Budapest, Hungary
e-mail: bede.zsuzsanna@mail.bme.hu

P. Gáspár
Institute for Computer Science and Control, Hungarian Academy of Sciences,
Budapest, Hungary
e-mail: gaspar.peter@sztaki.mta.hu

M.C. Tandon and P. Ghosh (eds.), *Mobility Engineering*,
DOI 10.1007/978-981-10-3099-4_14

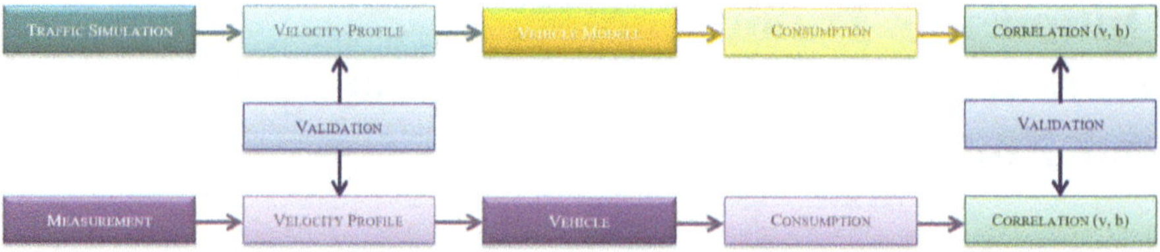

Fig. 14.1 Simulation versus measurement

smallest travel time, least fuel used, lowest emission of pollutants) based upon the above data.

2 The Validation of the Two Kinds of Models

The aim of the research is to analyse the interconnection of the two kinds of models (vehicle and traffic network) from the aspect of a traffic network model [3]. We can control a vehicle by designing its speed trajectory. Turning this around, we can say that by controlling the traffic we can influence the speed of the vehicles. Thus, the aim of the control is to improve the vital qualities of all vehicles in the given junctions.

The validation of the model is based on the comparison of data obtained from the model describing the network system and from the model describing the unique vehicle with vehicle dynamics parameters (Fig. 14.1).

The result of the joint analysis of the data obtained from the models and measured parameters is that the speed profile and consumption data can be considered homogeneous.

3 Conclusions

The primary aim of the emission-based traffic control is to achieve sustainable transportation with the reduction of emission, noise level and/or consumption by using the most significant effects of the emission (e.g.: average speed, composition of vehicle, intensity of a traffic flow, etc.). These effects were taken into account with different weights. With the help of this model, an adaptive environmental-based traffic control is realizable. (The adaptive traffic control is a classical control loop with measurements and/or estimates, control and interrupts.)

References

1. A. Mihály, B. Németh, and P. Gáspár, “Analysis of driver behavior related to look-ahead control,” 13th IFAC Symposium on Control in Transportation Systems, pp. 268–273, 2012.
2. B. Németh and P. Gáspár, “Design of vehicle cruise control using road inclinations,” International Journal of Vehicle Autonomous Systems, vol. 11, no. 4, pp. 313–333, 2013.
3. A. Mihály, B. Németh, Zs. Bede and P. Gáspár, “Look-ahead cruise control design in VISSIM simulation environment” Models and Technologies for Intelligent Transportation Systems (MT-ITS) 3–5. June 2015. Budapest, Hungary, pp. 52–57, 2015.

Author Index

M.C. Tandon and P. Ghosh (eds.), *Mobility Engineering*,
DOI 10.1007/978-981-10-3099-4

CPI Antony Rowe
Chippenham, UK
2018-06-25 22:21